D0216667

Enhancing Evolution

Enhancing Evolution

The Ethical Case for Making Better People

John Harris

PRINCETON UNIVERSITY PRESS | *Princeton and Oxford*

Published by Princeton University Press,
41 William Street, Princeton, New Jersey 08540

In the United Kingdom: Princeton University Press,
3 Market Place, Woodstock, Oxfordshire OX20 1SY

ISBN: 978-0-691-12844-3 (alk. paper)

Library of Congress Control Number: 2007928679

A catalogue record for this book is
available from the British Library

This book has been composed in Palatino

Typeset by T&T Productions Ltd, London

Printed on acid-free paper ∞

press.princeton.edu

Printed in the United States of America

10 9 8 7 6 5 4 3 2 1

For Jacob

Contents

Foreword

By Steve Rayner[1]

This book evolved from a series of invited public lectures delivered at the James Martin Institute for Science and Civilization. The institute was founded at Oxford University's Saïd Business School in 2004, supported by a generous benefaction from the James Martin Trust.[2] Our mission at the institute is to focus on the major science and technology issues that are likely to shape the next decade to century. Where possible, we seek to highlight opportunities where relatively modest investments or interventions in the present have the potential to propagate, over time, to prevent harms and encourage the generation of greater goods for humanity.

To this end, the institute is developing two core capabilities. The first is in long-range thinking about alternative possible states of the world, although not necessarily all of it at the same time. We seek to think intelligently and nondeterministically about the future of a wide range of institutions, including families, communities, businesses, local and national governments, and international agencies. The second capability lies in understanding human behavior as complex adaptive systems, subject to enormous challenges of path dependency (technological and social lock-in effects), such as those characterizing the world's current commitment to fossil fuels and consequent harmful greenhouse-gas emissions. Our goal in such cases is to understand how heavily overdetermined systems can be unlocked, to facilitate institutional and technological change, especially in the face of entrenched interests and commitments to competing values.

We seek to bring these twin capacities, which we refer to as "futures" and "practices" to bear in six areas of application:

- tomorrow's technologies—understanding the social issues around new and emerging technologies, such as nanotechnology and biotechnology, as well as the implications of their convergence;
- governance of technological change—recognizing that twenty-first-century technologies are globally pervasive from their emergence and that the days of national or regional experimentation are past;
- tomorrow's planet—focussing on the large, tightly coupled technological and natural systems, such as energy and climate or water, agriculture, and energy;
- technology and inequality—facing the challenge of creating opportunities for technology to ameliorate, rather than exacerbate, social and economic inequality;
- tomorrow's civilization—exploring issues of community identity, and the roles of citizens and consumers in a global technological society;
- tomorrow's people—investigating the implications of radical changes in health technology and the potential for radical life extension and human capacity enhancement.

When Princeton University Press invited us to propose a series of lectures on a subject close to the institute's heart, we had no hesitation in selecting the sixth theme as an exciting and provocative topic, with implications for several of the others. There was immediate consensus on the individual we would invite to deliver the lectures on three successive evenings in March 2006, coinciding with the first James Martin Institute World Forum on Science and Civilization, which tackled the same broad topic. Professor John Harris is not only a distinguished academic, but an engaged and engaging public-policy commentator and advisor, well-known to the public through his many broadcasts on television and radio. We were very gratified by his enthusiasm for the project.

The resulting book speaks for itself with characteristic clarity and forthrightness. It is, of course, a work of philosophy and is committed to an ethical position located in the liberal, democratic, utilitarian tradition, and broadly in favor of both extending the human lifespan and embracing human capacity enhancements both physical and intellectual. Professor Harris's approach to life extension and enhancement is generally permissive. Therefore, his presumption is that such technologies should be pursued unless someone comes up with a compelling reason not to. His engagement with diverse arguments is both cogent and comprehensive. He provides us with a view of the intellectual landscape at the same time as taking on the hard arguments of other commentators who have expressed strong reservations or even outright opposition to "transhumanist" applications of medical technology.

In doing so, he touches on several themes that are of more general interest to the broad range of topics to which the institute addresses itself. For example, in relation to *tomorrow's technologies*, some commentators (both for and against) represent technologies of life extension and enhancement as novel, revolutionary breakthroughs liable to turn the world upside down for good or ill. This book tends to view extension and enhancement technologies as incremental extensions of existing ones, such as eyeglasses, binoculars, and even writing. Elsewhere, I have described the tension between claims of technological discontinuity and continuity in policy debates as the "novelty trap," and David Edgerton addresses it in a recent book, *The Shock of the Old*.[3] However, the same tension seems to be a consistent feature of the emergence of new technological fields, from nuclear power in the 1970s to nanotechnology in the present decade. Claims of novelty generate excitement and funding, but can also prompt concerns about novel, unanticipated risks—what Harris, in his lecture, called the shift from "wow to yuck." This tension is clearly an important issue for our studies of *tomorrow's technologies* and *governance of technological change*.

Professor Harris also raises another significant issue for the *governance of technological change* in the specific context of embryo selection. This is an area where the United Kingdom's Human Fertilisation and Embryology Authority (HFEA) consulted widely with the public prior to deciding against allowing sex selection of embryos, except to prevent transmission of serious genetic disorders. He proposes that the United

Kingdom could issue a significant but limited number of sex-selection licenses and monitor the effects that it has on society, without running the risk of major disruption to either the demographic or moral order. This kind of experiment would represent an interesting opportunity to study what others have called "real-time technology assessment"[4] as a practical alternative to a stark choice between the "proof-first" approach to regulation, which places the burden on the regulator to demonstrate a real danger of specific harm, and the "precautionary principle," which counsels "if in doubt, don't." Real-time technology assessment seems to offer a viable path that would allow for the monitoring and real-time regulation of emerging technologies and that could be superior to the stale and unproductive choice between stultifying moratoriums, which do not allow for social learning, and those who would recklessly plough "full steam ahead and damn the torpedoes," thereby exposing society to unanticipated negative consequences downstream. Indeed, the idea behind the HFEA itself could be viewed as a limited experiment in real-time technology assessment.

The sex selection case also raises interesting questions about the current enthusiasm for increasing direct public participation in technological decision making. By Harris's account at least, the public was poorly informed about the issues of sex selection, but overwhelmingly opposed, largely on the "yuck" principle. There is a clear warning here that increased public engagement must mean engagement with a plurality of viewpoints, not just the expression of popular prejudice if, indeed, it is going to become a regular feature of our modern democracy.

Elsewhere in this volume, Professor Harris devotes considerable attention to another core Martin Institute concern: the issue of *technology and inequality*. A major worry of many critics is the possibility that unequal access to new enhancement technologies will exacerbate existing gaps between rich and poor. Some have even suggested an accelerating danger of speciation as humanity divides into the "enhanced" and the "naturals." Professor Harris points out that, historically, it has been the ability of elites to access innovations that has led to new knowledge or technology eventually becoming available to the masses. After all, if the medieval scholars of Oxford had abjured writing as elitist, it is unlikely that aspirations for universal literacy would be the norm today. While trickle-down theories seem to have historical power, we must also recall that there were past efforts to limit literacy and access to the written word (chained bibles, for instance) and we cannot be complacent about the issues raised by access to technology as part of what Amartya Sen would describe as the "entitlements" of economic and social development.[5]

A distinctive characteristic of the Martin Institute's approach to challenges of the twenty-first century is a focus on what we call "wicked problems, uncomfortable knowledge, and clumsy solutions." Observing that, in the normal course of events, three to five embryos perish for each live human birth, John Harris concludes that

> [w]e, humankind, must accept that human embryos are deeply ambiguous and problematic entities of a kind whose lives or "dignity" simply cannot be protected in ways consistent either with other values that we hold or indeed with the continued existence of the human species. The alternative is consistent but bleak, it involves the strict avoidance of all acts which would violate the sanctity of life of embryos. This would of course include almost all human procreation and certainly all sexual reproduction.

Such knowledge is deeply uncomfortable for those who are committed to the idea that the sanctity of human life begins at the moment of conception. It highlights what some have called the "wickedness" of problems that face human societies in which diverse constituencies are deeply committed to often contradictory certitudes about public policy issues.[6] We know that the policy pathways for dealing with them will often be meandering, providing complete satisfaction for no one, but hopefully arriving at "clumsy solutions" which avoid the deepest offence to most.[7] These "wicked problems" characterized by "uncomfortable knowledge" and susceptible (in the absence of dictatorship) only to "clumsy solutions" are exactly the kinds of problem that the Martin Institute seeks to identify and to cast light on, for scholars, politicians, businesses, and citizens.

I offer these as just a sample of the many issues that this book raises which have reverberations far beyond the particular subject of life extension and enhancement. The book is therefore exciting both as a piece of specific scholarship and for its insights into a broader range of challenges for the twenty-first century. We are confident that this book will be the first of many important, illuminating, and provocative volumes arising from the James Martin Institute Lectures.

Notes

1. James Martin Professor of Science and Civilization and Director, James Martin Institute for Science and Civilization, University of Oxford.
2. In 2005 it also became a core component of the University's James Martin 21st Century School, which was established by a further benefaction from the trust.
3. Rayner, S. 2004. The novelty trap: why does institutional learning about new technologies seem so difficult? *Industry and Higher Education* 18:349–55. Edgerton, D. 2006. *The Shock of the Old: Technology and Global History Since* 1900. Oxford University Press.
4. Guston, D., and D. Sarewitz. 2002. Real-time technology assessment. *Technology in Society* 24:93–109.
5. Amartya Sen has addressed the issue of poverty and entitlement in a series of works spanning at least three decades.
6. Rittel, H., and M. Webber. 1973. Dilemmas in a general theory of planning. *Policy Sciences* 4:155–69.
7. Verweij, M., and M. Thompson (eds). 2006. *Clumsy solutions for a complex world.* Basingstoke: Palgrave.

Acknowledgments

Many people have helped and influenced me in the writing of this book. I am indebted to the James Martin Institute of Science and Civilization for inviting me to present "The Princeton Lectures" as part of the "First World Forum on Science and Civilization" at the Saïd Business School, University of Oxford, March 14–17, 2006. Equally my debt is to Princeton University Press for sponsoring these lectures and for commissioning me to write this book arising from the theme of the World Forum: "Tomorrow's People: The Challenges of Technologies for Life Extension and Enhancement." I learnt much from the colleagues who attended the World Forum and from the wonderful audiences who shared their ideas with all of us. Over many years I have had conversations related to these themes with a number of people. In particular I wish to record thanks to Katrien Devolder, Søren Holm, Simona Giordano, Allen Buchanan, Dan Brock, Stephen Minger, John Sulston, Gordon Jason, Nancy Rothwell, Jacob Harris, Daniela Cutas, Muireann Quigley, Dan Wikler, Sarah Chan, Fu Chang Tsai, Jay Olshansky, Tom Kirkwood, Aubrey De Grey, Sally Greengross, Ray Tallis, Steven Rose, Antonia Cronin, Margaret Brazier, Simon Winner, Julian Savulescu, Steve Rayner, Amel Algrahni, Sarah Devaney, James Martin, Lisa Bortolotti, and Sita Williams.

A special debt is owed to Nir Eyal and Julian Savulescu for services way beyond the call of duty in providing very detailed comments on an earlier draft of this book. Their intellectual rigor pervades this book and if I have not recorded each debt I owe them, it is only in order to save the reader tedious repetition.

Parts of this book have been used elsewhere and where this has happened there are references in the text. In one or two places I have drawn on work I published with coauthors; again references will reaffirm the debts I owe. In particular I thank Søren Holm and Katrien Devolder for permission to quote from our joint efforts.

I wish to record debts to two philosophical mentors and friends who have shaped my ideas and been a constant example to me of how philosophy should be done: with humor, rigorous intellectual honesty, and devastating effect. They are Ronnie Dworkin and Jonathan Glover. Three other "mentors" whom I never met, but whose wit and brilliance are always before my mind, are Bertrand Russell, Douglas Adams, and F. M. Cornford.

Other books on this subject have undoubtedly affected the way that I think about these issues, but I have not revisited them in preparing this book. These include, first and foremost, Jonathan Glover, *What Sort of People Should There Be?* (Pelican, Penguin Books, 1984). Since I wrote my first book on human enhancement (*Wonderwoman and Superman*, Oxford University Press, 1992), some of the principal texts which have taken up related themes are Philip Kitcher, *The Lives to Come* (Simon & Schuster, 1996), Lee Silver, *Remaking Eden* (Weidenfeld & Nicolson, 1998), and Gregory Stock, *Redesigning Humans* (Houghton Mifflin, 2002).

Debts of a different kind are owed to the European Commission for funding three projects, EURECA, CLEMIT, and EUROBESE, which have supported my research on this book.

Daniela Cutas has read the entire manuscript and made many invaluable suggestions and saved me from many errors. Richard Baggaley, my editor at Princeton University Press, has been a constant support, and Emma Dain of T&T Productions Ltd, London, not only edited the electronic files in exemplary fashion, but made many valuable suggestions. Thanks to her the editing process has been a genuine pleasure. It is perhaps therefore somewhat surprising that I still have the effrontery to claim authorship and acknowledge that I am responsible for any remaining errors.

Enhancing Evolution

Introduction

From "Yuck!" to "Wow!" and How to Get There Rationally

Suppose a school were to set out deliberately to improve the mental and physical capacities of its students. Suppose its stated aims were to ensure that the pupils left the school not only more intelligent, healthier, and more physically fit than when they arrived, but more intelligent, healthier, and more physically fit than they would be at any other school. Suppose they further claimed not only that they could achieve this but that their students would be more intelligent, enjoy better health and longer life, and be more physically and mentally alert than any children in history. Suppose that a group of educationalists, outstanding ones of course, far more brilliant than any we know of to date, had actually worked out a method of achieving this, in the form perhaps of an educational and physical curriculum. What should our reaction be?

Well of course our reaction would be one of amazement; it would certainly be an unprecedented event—a breakthrough in education. But should we be pleased? Should we welcome such a breakthrough? We might of course be skeptical, we might doubt such extravagant claims, but if they could be sustained would we want our children to go to such a school? And if the school our own children attended was not run according to the new educational methods, would we want these to be adopted as soon as possible?

We ought to want this. It is, after all, part of what education is supposed to be for. Indeed, if the claims were not expressed hyperbolically and competitively, this is what, if we knew little about education, we might well imagine was actually going on in schools or at least was what the teachers were supposedly trying to bring about. Of course we might have some reservations. We might want to be assured that others of the things we want from education would not be sacrificed in the cause of intelligence and bodily health. We would want this school and these educational methods to transmit our culture, and we would want our children prepared for the real world. But, this said, if the gains in intelligence, fitness, and health were significant and palpable we might well be willing to postpone initiation into some elements of our multicultural heritage or forgo some extra periods of personal and social education for the sake of these compensating gains in health, intelligence, and physical fitness.

Now suppose, as is much more likely, we could use genetic engineering, regenerative medicine or drugs, or reproductive technology or nanotechnology to produce healthier, fitter, and more intelligent individuals. What should our reaction be? Would it be unethical to do so? Would it be ethical not to do so?

Our question is this: if the goal of enhanced intelligence, increased powers and capacities, and better health is something that we might strive to produce through education, including of course the more general health education of the community, why should we not produce these goals, if we can do so safely, through enhancement technologies or procedures?

If these are legitimate aims of education, could they be illegitimate as the aims of medical or life science, as opposed to educational science?

Enhancements of course are good if and only if those things we call enhancements do good, and make us better, not perhaps simply by curing or ameliorating our ills, but because they make us better people. Enhancements will be enhancements properly so-called if they make us better at doing some of the things we want to do, better at experiencing the world through all of the senses, better at assimilating and processing what we experience, better at remembering and understanding things, stronger, more competent, more of everything we want to be. A welcome part of all this added value is the likelihood, the hope, and intention, that enhancements will also make us less: less the slaves to illness, pain, disability, and premature death; less fearful because we have less to fear; less dependent, not least upon medical science and on doctors.

For these and many other reasons which we will examine as we proceed, this book defends human enhancement and argues that not only are enhancements permissible but that in some cases there is a positive moral duty to enhance.

If there is a theme which unites all my philosophical work, it is an exploration of the responsibility shared by all moral agents, to make the world a better place. Karl Marx[1] is noted for the idea that the purpose of philosophy cannot simply be to understand the world, but must also be to change it. This thought, however, is not original to Marx; it is implicit in the writings of many philosophers—Plato certainly wanted to change the world for the better and *The Republic* is devoted to systematic ways to achieve a better society. Locke, Rousseau, and Bentham would all have been equally at home with the idea. Indeed, as Bertrand Russell said, talking of Jeremy Bentham:

> There can be no doubt that nine-tenths of the people living in England in the latter part of the last century were happier than they would have been if he had never lived. So shallow was his philosophy that he would have regarded this as a vindication of his activities.[2]

Russell's irony will not be lost on even the most literal of readers. It is a sad comment on the philosophy of the twentieth century that in the four score years since Russell's essay was written, concern with the real world, no less than with attempts to make it better, have continued to be seen as evidence of lack of philosophical depth by the majority of professional philosophers, and Russell's own attempts to make the world better are not, even now, ranked by most philosophers as among his significant philosophical contributions. All these philosophers place philosophy at the service of humanity, for what use is knowledge and understanding without using that understanding to try to change things for the better?

It is significant that we have reached a point in human history at which further attempts to make the world a better place will have to include not only changes to the world, but also changes to humanity, perhaps with the consequence that we, or our descendants, will cease to be human in the sense in which we now understand that idea. This possibility of a new phase of evolution in which Darwinian evolution, by natural selection, will be replaced by a deliberately chosen process of selection, the results of which, instead of having to wait the millions of years over which Darwinian evolutionary change has taken place, will be seen and felt almost immediately. This new process of evolutionary change will replace *natural selection* with *deliberate selection, Darwinian evolution* with "*enhancement evolution*."

One of the ways in which philosophy can contribute to a better world is to help clear away the bad arguments that stand as much in the way of human progress and human happiness as do reactionary forces of a political and even of a military kind. When new technologies are announced, the first reaction is often either "wow—this is amazing!" or "yuck—this is sick!" This book is about the reasons and arguments that underlie both reactions, and about how it can sometimes be rational to move from "yuck!" to "wow!" In the chapters that follow I present the arguments for human enhancement and analyse human enhancement; the book builds on work of the last twenty years which has at its center the moral responsibility of human beings to make responsible, informed choices about their own fate and the fate of the world in which we live. In the face of threats both to humankind and indeed to the ecosystem which sustains us and all life, this responsibility is nothing short of a clear imperative to make the world a better place.

This is a book of arguments; in the course of it I try to critically evaluate the main arguments opposing human enhancement of all forms and at the same time, through use of real examples and discussion of present and future enhancement technologies, I develop the arguments and the good reasons we have not only to take the possibility of radical human enhancement seriously but also positively to promote it. I point out the continuity that exists between therapy and enhancement, the fact the human enhancement has always been both a conscious and unconscious part of human development and of evolution, and I underline the familiarity of the multifarious attempts we humans have made not only to better our ourselves in the sense of improving our material circumstances and well-being, but literally to better ourselves. In short, I propose both the wisdom and the necessity of intervening in what has been called the natural lottery of life, to improve things by taking control of evolution and our future development to the point, and indeed beyond the point, where we humans will have changed, perhaps into a new and certainly into a better species altogether.

As the argument develops, a radical thesis outlining the moral, political, and social reasons to welcome the prospect of enhancement is developed, as is a defense of the idea of making people, or rather permitting people to make themselves and their children, longer-lived, stronger, happier, smarter, fairer (in the aesthetic and in the ethical sense of that term) and in finding ways to do this which will protect the safety of the people and of course be consistent with good government and regulation.

The Agenda

In the first three chapters I argue that the opportunity to create healthier, longer-lived, and altogether "better" individuals is one that there are moral reasons to take and that it is an opportunity that is in the interests of the individual, society, and government. Indeed, governments have prudential as well as moral reasons to support parental and individual choice in such matters. It is further argued that the freedom of citizens to do what's right ethically and what's right for them personally is not only self-evidently sensible, it is enshrined in our moral and political theory.

Chapter 1 is really a further introduction to and explanation of the themes of this book, but it seeks to do more than to pose questions or to set out the conclusions that are coming: it also begins to argue for these conclusions. Chapter 1 also further explains my own commitment to making the world a better place and how I see philosophy and bioethics as rational ways of attempting to do so.

The three opening chapters examine the various techniques that might be used for enhancement and the various targets of those techniques. Stem cell research and therapy, gene manipulation, selection of embryos, drugs, machines and other mechanical enhancements, and many other enhancement techniques are considered and evaluated. These chapters also advance a new thesis as to how health and disease are to be understood, which replaces and shows the unacceptability of the previous model advanced by Boorse and Daniels.

Chapter 4 considers in detail perhaps the most radical and far-reaching of all types of enhancement, that of the possibility of life extension, perhaps to the point where the enhanced individuals would for all practical purposes be immortal. The arguments for and against such a dramatic possibility are considered in detail in the light of the fact that life extension is simply the corollary of lifesaving and is therefore an established part of our common-sense morality.

Chapter 5 looks at reproductive choices as a way of influencing the sorts of people that will be brought to birth and exist in the future. The many arguments which purport to give reasons for limiting access to reproductive technologies and procedures which may facilitate the enhancement of individuals or permit illness, impairment or disability to be removed or minimized are examined. A crucial issue here is whether or not these arguments point to dangers or harms of sufficient seriousness or sufficient probability or proximity to justify the limitation on human freedom that they require. I argue that there is a so-called "democratic presumption" in favor of freedom and in particular supporting freedom of reproductive choice. The presumption is that citizens should be free to make their own choices in the light of their own values, whether or not these choices and values are acceptable to the majority. Only serious real and present danger, either to other citizens or to society, is sufficient to rebut this presumption. The question of whether or not such serious dangers attend freedom of reproductive choice is considered in depth.

Chapter 6 examines disability and disadvantage and asks whether attempts at human enhancement constitute some sort of insult to people with disabilities or disadvantages and whether or not attempts to make better people constitute unfair discrimination against people with disabilities. Might increases in the gap between normal and enhanced abilities, as well as between enhanced and disabled people, be a decisive objection to enhancement?

Chapters 7 and 8 analyse in detail the work of three prominent and complex theorists, Michael Sandel, Leon Kass, and Jürgen Habermas, all of whom have mounted a vigorous and sustained critique of human enhancement and all of whom would stop attempts at enhancement in their tracks. I suggest that the arguments of all these thinkers are inadequate or misconceived as objections to enhancement.

In chapter 9 I will look at enhancements which are not principally focused on health or treatment but are more "cosmetic" or "elective."

The choice of phenotypical traits such as hair, eye and skin color, physique, stature, and gender are examples of what I call morally neutral choices in the sense that it is not in most circumstances better or worse, morally speaking, to be black, white, tall, short, male or female, brown-haired or blond (despite what gentlemen allegedly prefer). Sex selection is taken as a case study and both the ethics and the policy dimensions of permitting such choices are explored.

In the final two chapters I focus on the research that will be required if new forms of enhancement are to be developed and made available. The penultimate chapter takes as its point of departure the fact that many of the most dramatic forms of enhancement will use or will be a by-product of therapies and techniques using regenerative medicine and stem cell science to achieve both the therapeutic and the enhancement effects. The ethics of these techniques will, for the foreseeable future, turn on the legitimacy of sourcing stem cells from embryos and indeed on research using embryos or embryo-like entities. The moral status of the embryo is therefore of vital importance to the possibility of human enhancement, at least for the foreseeable future. This chapter is not a traditional examination of the familiar problem of "moral status."[3] Rather, I argue that the embryo is an irredeemably ambiguous entity and that human life, and even probably post-human life, is simply not possible for creatures that regard the embryonic forms of their species as sacred.

Finally I turn to the role of science research in the contemporary world. I argue that research is so essential a part of life, society, and the possibility of human progress that a radical reevaluation of the role of research in our lives is required. Hitherto, research has been viewed with suspicion and regarded as a sort of "optional extra." I argue that if human life and welfare are to continue to be protected and the enhancements that offer the best prospects for future health and welfare are to be developed, not only must research be valued but support of, and indeed participation in, research must in some circumstances be regarded not only as desirable but as a positive moral obligation.

1 | Has Humankind a Future?

> *"Se vogliamo che tutto rimanga come è, bisogna che tutto cambi."*
>
> *"If we want things to stay as they are, things will have to change."*
>
> —Tomasi de Lampedusa, *The Leopard*

Wouldn't it be wonderful if we humans could live longer healthier lives with immunity to many of the diseases like cancer and HIV/AIDS that currently beset us? Even more wonderful might be the possibility of increased mental powers, powers of memory, reasoning, and concentration, or the possibility of increased physical powers, strength, stamina, endurance, speed of reaction, and the like. Wouldn't it be wonderful?

Many people think not. The idea of improving on human nature has been widely rejected. Decisive interventions in the natural lottery of life, to enhance human performance, improve life, and perhaps thereby irrevocably to change our genetic constitution, have met with extreme hostility. This hostility is, as we shall see, misplaced. In this book I hope to convince you that human enhancement is a good thing and that our genetic heritage is much in need of improvement.

Whatever people say, no one, I believe, actually thinks that there is anything in principle wrong with the enhancement of human beings. This seeming contradiction, paradoxical as it may appear, is resolved when we reflect on the familiarity and acceptability of existing enhancement technologies and on their history. Many of us are already enhanced (do you wear glasses, for example?) and all of us without exception have benefited from enhancing technologies. (For example, have you ever been immunized? And even if you haven't, you will have benefited from the so-called "herd immunity" created by the fact that others have.)

Not only do we all approve of enhancement, we approve for good reasons—we approve because we are decent moral people who want to protect each other from harm and who want to benefit ourselves, and others.

In terms of human functioning, an enhancement is by definition an improvement on what went before. If it wasn't good for you, it wouldn't be enhancement. There is a continuum between harms and benefits and the reasons we have to avoid harming others or creating others who will be born in a harmed state are continuous with the reasons we have for conferring benefits on others if we can.[1]

We have reasons for declining to create or confer even trivial harms and we have reasons to confer and not withhold even small benefits.[2] The opportunity to create healthier, longer-lived, and altogether "better" individuals is one that there are moral reasons to take.

As with all opportunities, we have also to consider the risks that they may entail and there is of course a relation between the magnitude and probability of the benefit and the degree and size of risk we are prepared to run to get it.

I will argue that enhancement is also an opportunity that it is in the interests of society and government to take. On this view, parents would act ethically if they were to attempt to achieve such an objective for their children, and those of us who are autonomous enough to consider such questions have good reasons to confer such benefits on ourselves.[3] I will further show that governments have prudential as well as moral reasons to support parental and individual choice in such matters. Indeed, although this chapter is partially intended to introduce the themes of this book, it also initiates the argument and attempts to place the argument in a tradition of thinking about attempts to shape human nature and to rethink the destiny of humankind.

The freedom of citizens to do what's right ethically and what's personally prudent is not only self-evidently sensible, it is, as we shall see, enshrined in our moral and political theory. Now, and in the chapters that follow, I will show why and how human enhancement is in the interests of all of us personally and in the interests of society. The principal objections to human enhancement will be examined in detail and I will argue strongly not only for the freedom, but also for the obligation to pursue human enhancement.

Threats to human life and dramatic policies and practices to meet them are all too frequent in human history. My own interests in this process began in the early sixties of the last century.

In 1961 the philosopher Bertrand Russell published a book (a pamphlet really) asking a pertinent and agonizing question. That question had arisen because of threats posed by scientific advance and humankind's apparent inability to deal sensibly with the consequences. In 1961 the perceived threat was to all human life and it came from the policies of "mutually assured destruction" which were at the heart of strategies on both sides in the cold war concerning the use of nuclear weapons. Thinking about a rational response to what he perceived as the real possibility of the extinction of all human life, Russell's book asked the question and took the title: *Has Man a Future?* This book asks effectively the same question, and seeks to examine whether the answer might not lie in humankind's ability to realize its potentialities. We should be clear that, while the question "has humankind a future?" seems to be empirical, this is not the case. The question invites reflection on the nature of humankind and on the desirability of humankind's continued existence or further evolution.

Russell imagines a conversation with God:

> If I were the pleader to Osiris for the continuation of the human race, I should say: "O just and inexorable judge, the indictment of my species is all too well deserved, and never more so than in the present day. But we are not all guilty and few of us are without better potentialities than those that our circumstances have developed.... It is not only what to avoid that great men have shown us. They have shown us also that it is within human power to create a world of shining beauty and transcendent glory.... Lord Osiris we beseech thee to grant us a respite, a chance to emerge from ancient folly into a world of light and love and loveliness."
>
> Perhaps our prayer will be heard. In any case, it is because of such possibilities, which, so far as we know, exist only for Man, that our species is worth preserving.[4]

Russell was concerned with a nuclear catastrophe that might destroy all life and end in a permanent nuclear winter. He was pleading for the survival of the human species: not a survival that would preserve it with all its faults and follies, but rather a survival that would enable the realization of our "better potentialities." At the conclusion of *Has Man a Future?*, Russell restates his hopes for a future free from the fear of the threat that loomed largest in 1961:[5]

> Man has not only the corresponding capacities for cruelty and suffering, but also potentialities of greatness and splendour, realised, as yet very partially, but showing what life might be in a freer and happier world. If man will allow himself to grow to his full stature, what he may achieve is beyond our present capacity to imagine. Poverty, illness and loneliness could become rare misfortunes ... And with the progress of evolution, what is now the shining genius of an eminent few might become a common possession of the many. All this is possible, indeed probable, in the thousands of centuries that lie before us.[6]

Russell might initially be taken to be saying that the survival of humans as we know them has value. But if we scratch the surface we see that for him what matters is the preservation and expansion of what is good about humans. We too should prioritize improving on humans over preserving the species in its present form.[7]

This "progress of evolution" is unlikely now to be achieved accidentally or by letting nature take its course. If illness and poverty are indeed to become rare misfortunes, this is unlikely to occur by chance, even with the thousands of centuries that Russell envisages and evolution requires. It may be that a nudge or two is needed: nudges that will start the process, trailed in the introduction to this book, of replacing *natural selection* with *deliberate selection*, *Darwinian evolution* with "*enhancement evolution*." This book is concerned with the ethics and indeed with the policy dimensions of providing the required nudges.

If we wish humankind to achieve its potential (which has so far almost universally been assumed to be an inevitable part of evolutionary progress), this might require some deliberate changes. It is possible that, however conservative we are and however much we wish for things to go on as they are, things will have to change. If this is right, then conservatives for whom the sanctity of the existing human genome or the preservation of the species is an article of faith may need to accept change to preserve if not the totality, at least the essence, of what they value.

The epigraph for this chapter, which perhaps also serves as a salutary reminder of the connection between the impulse to conservatism and the temptations of revolutionary change, is Lampedusa's wonderfully memorable idea that "If we want things to stay as they are, things will have to change."[8] It is worth recalling the context in which this paradox is introduced.

Don Fabrizio, Prince of Salina, the "Leopard" of the title of Lampedusa's famous novel, is a deeply conservative hereditary prince in a still feudal Sicily. When we meet him in May 1860, Garibaldi's "revolution" is about to topple the Bourbon monarchy and the "Kingdom of the Two Sicilies." Ten years later, a united Italy, a phenomenon not seen since the fall of the Roman Empire, will be proclaimed, with Rome as its capital, and the Risorgimento will have completed its final phase. The Prince of Salina wants things to go on as they always have and yet he eventually accepts that his nephew Tancredi is right to think that even revolutionary change may sometimes be the only way to protect things as they are and that "if we want things to stay as they are, things will have to change."[9] The very different possibilities that are the subject of this book raise the same questions that faced Tancredi and Don Fabrizio: whether or not change is required and what will have to be sacrificed to achieve it, whether the changes constitute a revolution or the continuation of the status quo by other means. The crucial question, however, concerns whether or not the proposed changes enable the survival of what matters and permit the flourishing and further improvement of life for everyone (which is what all decent conservatives, liberals, and revolutionaries desire).

Russell's question is thus also the Prince of Salina's dilemma. It is a question and a dilemma that has been sharpened in our own times by the revolutionary possibilities for human enhancement developed and developing in science and technology.

This book looks seriously at the possibility of revolutionary change in human powers and capacities as well as in human nature. Whether this revolutionary change will prove conservative or radical is of course a further and complex question. Potentially, supporting human enhancement is as conservative as the Prince of Salina and as revolutionary as Tancredi; it shares with both of them their love of life and of love, their respect for science, for tradition, and acceptance of the necessity, sometimes, for decisive action.

It is doubtful that there was ever a time in which we ape-descended persons were not striving for enhancement, trying to do things better and to better ourselves.

Shelter, learning and teaching, tool using, body decoration, clothing, gathering and hunting, cooking, storing, cooperation, cultivation, animal taming and domestication, farming, social living, language, and education are all enhancement techniques or technologies. With the help of some of these tools we have built institutions and relationships, families, villages and cities, societies and civilizations, schools, universities, markets, commercial organizations, and other mechanisms of cooperation and competition. We have created literature, art, and music; we have created agriculture and industry, science and medicine, and technology and engineering.

Not all of these are equally beneficial of course, and any human construct can be misused or ill-used; but all mechanisms which make possible (though not of course inevitable) better life and better lives are means of enhancement in some sense. Substances that are effective analgesics in small doses are often poisons in larger doses; heroin is a derivative of morphine and both are derived from opium. The fact of widespread and disastrous heroin addiction does not discredit morphine as an important source of pain control and hence as a "beneficial" compound.

It is important to be clear that when we call something an "enhancement" or an "enhancing" technology or therapy we are not saying that is always its effect, any more than when we call something an "analgesic" we imply that it will in every case and every dose reduce pain or that "stimulants" will always stimulate or that "carers" always care or "healers" always heal. There is no sensible way in which we must take the possibility of misuse into account *before* determining that something is an enhancement. When we call something a form or method of human enhancement, we are pointing to a likely improvement that it can (will typically) effect if used in ways best calculated to achieve that effect. Aspirin is a painkiller which can also have many other beneficial "enhancing" effects, including reducing the risk of stroke. It is also a well-known killer and is often used to attempt suicide.

Writing is one of the most significant enhancement technologies.[10] In addition to providing ways of recording speech and setting down ideas, it supplements memory and has indeed made, among many other things, recorded history possible for the first time. Writing, and everything that goes with it—reading, books, printing, libraries, education, universities, computers, etc.—would never have taken off if its incredibly expensive and elitist beginnings had been like the serpent in the egg "killed in the shell."[11] Think how expensive, rare, and elitist were manuscript books, school and university education, and printed books. Remember too that, until comparatively recently, the ability to read, and in turn to teach literacy and numeracy and everything which depends on these, was very thinly spread, not only throughout the world but within most nation (and city) states.[12] Imagine if someone had said (and been heeded) that we should not invest in books and in literacy and education because it was expensive and elitist and could not be provided for all, or that we should not do so until it could be provided for all.

When the college at which I studied as a graduate student was founded in Oxford somewhere around 1263, university education was available to very few (and they were not the elite). The books the scholars read were in manuscript form and university education for all who might benefit was not even an idea, let alone an ideal. The same went for the founding of schools. Now in much of the world all children have some schooling, and university education is in many countries available for a majority of those who can (and who wish to) benefit. Schools and universities are incredibly expensive institutions but, despite this, it is not an exaggeration to say that the world aspires to universal provision. While this has not yet been achieved, few say that we should not further invest in education in Western Europe or North America, for example, until the same levels can be reached in the rest of the world. Equally we do not (and I believe should not) think that in investing in education we are trying to steal an advantage over people in countries who spend less.

If there is a lesson here, it is that we should be slow to assume that a good is too expensive, rare, or elitist to be pursued in the hope that eventually it can be made generally available, and that it therefore does not merit investment. Still less are there reasons to prevent the investment of others in the development of enhancing technologies and procedures.

Our collective origin as human beings occurred, almost certainly, in Africa between five and seven million years ago. Throughout the entire subsequent period we have been actively involved in enhancement, as well as passive or at least unwitting participants in an evolutionary process. To take one very simple example, every time we learn something new we change our brains: connections form in the brain which physically change its structure; these changes occur, if not permanently, then at least for a considerable period, and almost certainly improve our cognitive functioning.[13]

In a wonderful essay "Gaps in the mind,"[14] Richard Dawkins conducts the following thought experiment.

He asks us to imagine a contemporary woman (you or your sister) holding her mother's hand on the coast of Africa. She in turn holds *her* mother's hand, and she *her* mother's, and so on. Each daughter is as much like her mother as daughters usually are. Each person takes up a about a meter, a yard, of space as they hold hands back into the past. In just 300 miles (a small distance into Africa) the imaginary human chain reaches our common ape ancestor.

Now imagine our ape ancestor holding by her other hand *her* daughter, and she hers, and so on, back to the coast. Again each daughter looks as much like her mother as mothers and daughters usually do. By the time the chain reaches back to the coast, two contemporary females are looking at one another, each holding the hand of her mother stretching in seamless connection back to a common ape ancestor. The two "women," shall we call them, looking into each others' eyes are a modern human and a modern chimpanzee.

The point is not of course that we are descended from chimpanzees (we are not!); rather that both we humans and chimpanzees share a common ape ancestor, and we humans share our genetic origins with that ancestor and with chimpanzees. A mother shares 99.95% of her genes with her daughter but 99.90% of her genes with any randomly chosen person on the planet. Dawkins's story is a salutary reminder of the dangers of making a fetish of a particular evolutionary stage. If our ape ancestor had thought about it, she might have taken the view adopted by so many of our contemporary gurus, Leon Kass, Michael Sandel, George Annas, Francis Fukuyama, and many others, that there is something special about themselves and that their particular sort of being is not only worth preserving in perpetuity, but that there is a duty not only to ensure that preservation, but to make sure that neither natural selection nor deliberate choice permit the development of any better sort of being.

There are also those, and they are many,[15] who think that there are moral reasons to preserve not only human nature broadly conceived but also the human genome. They want it kept just as it is and regard it as a sort of genocide to think of further evolution into creatures that may no longer be human in the senses in which we understand the term. These conservatives may be like our imagined ape ancestor who (we have recklessly speculated) might have thought evolution had gone far enough and that her kind were at the very summit of imaginable evolution.

I personally am pleased that our ape ancestor lacked either the power or the imagination, or indeed avoided the errors of logic and/or morality, which might have led her to preserve herself at our expense. I hope that we will have the imagination, the power, and the courage to do better for ourselves and our descendants than the combination of chance, genes, and environment has done for us. This book is about the rights and wrongs of (and reasons for trying to) do just that. In the pages that follow I will not only set out the moral case for enhancement but indicate many of the ways in which enhancements can be ethically achieved.

I have emphasized the continuity of human enhancement and the fact that it has been part of human history from our first beginnings. This is an important reminder because it draws attention to our familiarity with the phenomenon of enhancement, the issues it raises, and the way in which it has changed and continues to change both ourselves and the world in which we live. This book is about the ethics and policy dimensions of human enhancement that have been occasioned by a new wave of technologies and possibilities that promise radical and different forms of enhancement than many of those with which we are familiar. It is also about contemporary fear of and reaction to these possibilities. This is my second book discussing this subject; the earlier work[16] was a comprehensive look at the possibilities that developing technology, particularly biotechnology, *might* open for human beings. That book was largely speculative; the technologies then envisaged were in their infancy and very little that today would be classified as the new wave of human enhancement was then possible.[17] Now, with a new wave of technologies, we also have renewed interest in and commentary upon human enhancement. There is also a new urgency, perhaps even a sense of panic in some quarters, now that possibilities formerly merely speculative are becoming a reality. In this book I will not attempt an exhaustive catalogue of the new ways in which human enhancement might be possible. I will, however, examine a sufficient range of the most likely avenues human enhancement might take to provide a comprehensive discussion of the ethics and policy issues they raise, and to try to resolve the most acute ethical dilemmas they pose. To do so, I will look not only at some of the most interesting and dramatic technologies, but at many of the most prominent reactions to them.

Many of the adverse reactions to human enhancement, as we shall see, take the form of dire predictions of the disasters that might attend on any attempts to change humans or human nature for the better. Of course we should take such possibilities seriously; dangers which might attend attempts at enhancement could wipe out any benefits and might indeed change things for the worse rather than for the better. However, unless we can see clearly how probable and serious the dangers are and have a realistic basis for balancing them against the probability and size of the benefits, we can have no rational basis for either precaution or enthusiasm. We will discuss the difficulty of such calculations in due course. I shall, however, suggest the appropriateness of a skeptical approach to such claims, demanding always an account of just what is allegedly harmful about the proposed course, rigorous assessment of the probability of such harms accruing, and a realistic basis for determining also their proximity. In short, we must know what reason there is to believe that it might be true that enhancements might be dangerous or indeed to think such a thing even plausible.

Again I am reminded of Russell's supremely rational approach to such dilemmas. In the opening paragraph of his book *Sceptical Essays*, he proposes a radical idea:

> I wish to propose for the reader's favourable consideration a doctrine which may, I fear, appear wildly paradoxical and subversive. The doctrine in question is this: that it is undesirable to believe a proposition when there is no ground whatever for supposing it true. I must, of course, admit that if such an opinion became common it would completely transform our social life and our political system; since both are at present faultless, this must weigh against it. I am also aware (what is more serious) that it would tend to diminish the incomes of clairvoyants, bookmakers, bishops and others who live on the irrational hopes of those who have done nothing to deserve good fortune here or hereafter.[18]

With this radical proposal in mind we must now turn to the very complex question as to whether the proposition that enhancements are a threat to humankind is true, and, if it is, as to whether there are good reasons to believe that ceasing to be human is in any way problematic.

2 | Enhancement Is a Moral Duty

I want now to make a brief but I hope decisive and persuasive case for the ethical imperatives which have placed human enhancement firmly on the agenda of all who care about the future of humankind. In later chapters I will deepen and broaden the arguments needed to defend and elaborate this thesis and consider in more detail the consequences of its acceptance. Here, the discussion is designed to introduce different types of enhancement defined by their modes of operation which give a vivid sense of the various forms the debate about enhancement takes. Different styles of objection to them and possible responses to those objections are "curtain raisers" to the more detailed discussion to come in later chapters.

Example 1: Mechanical versus Chemical Enhancements

I wonder how many readers of this book, like me, use spectacles?[1] All who do are using an enhancement technology. Now you might say "yes, but that restores normal functioning or repairs or corrects disease, damage, or injury." So it does.

Those who say this will probably know of the work of Boorse[2] and Daniels[3], who have each defined health and, hence, illness in terms of departures from normal functioning or departures from species-typical functioning.

Now consider the use of a telescope or pair of binoculars or a microscope. These tools are not used to restore normalcy or treat disease or injury. They are done to enhance powers and capacities.

Again I wonder how many of those who have ever used binoculars thought they were crossing a moral divide when they did so? How many people thought (or now think) that there is a moral difference between wearing reading glasses and looking through opera glasses? That one is permissible and the other wicked?

Those who think there is something in principle problematic about enhancement will have to show that there is some principled difference between spectacles and binoculars and that in the legendary incident at the battle of Copenhagen, England's great naval hero Horatio Nelson put the telescope to his blind eye to protect himself from moral turpitude![4]

Some people think that whether the enhancement is mechanical or chemical makes a moral difference. On this view bicycles are permitted but steroids not. Steven Rose, for example, makes this point forcefully. Having used substantially the same contrast as I have made between spectacles and telescopes, he goes on to say:

> It is true that when Galileo developed the telescope there were those among his compatriots who refused to look through it, but few today would share this ethical discomfort. Yet in the context of substances that interact directly with our bodily biochemistry, we feel a considerable unease, reflected in custom and law. It is alright to change our body chemistry by training, but to achieve a similar effect with steroids is illegal for athletes. It is alright to buy educational privilege for one's children by paying for private tuition, but dubious to enhance their skills by feeding them drugs.[5]

If we search in this passage (or elsewhere in Rose's writing) for a rational defense of this difference, we find only an appeal to custom and law or to the "yuck factor." "We feel," he says, "a considerable unease reflected in custom and law," and yet this same unease was felt by Galileo's compatriots with the same degree of justification! I will return to the distinction between mechanical and chemical enhancements in a moment, but first let's consider another example.

Example 2: Disease and Vaccination

Suppose there are some infectious diseases we can eliminate by operating on the environment. We can, we shall suppose, kill airborne infectious agents by introducing into the atmosphere a substance harmless to flora and fauna but lethal only to the target infectious agent, whether a bacterium or a virus. If we could be satisfied that the atmospheric additive was harmless to everything but the target virus, bacteria, or whatever, we would surely welcome such a discovery. Everything speaks for it and nothing against. Such an intervention would not, however, be an enhancement technology; it would not change human beings in any way. Rather, it would have the effect of rendering them safe from, if not immune to, these infections.

Now suppose we could eliminate the same infectious diseases by the use of effective vaccines in the way we have succeeded in doing with polio and smallpox. Presumably we would welcome this as a wonderful and effective public health measure, which saved lives, saved money, and minimized suffering and distress. Everything speaks for it and nothing against. Vaccination is of course an enhancement technology and one that has been long accepted (since the smallpox vaccine was first used at the end of the eighteenth century). Interestingly, there has been very little resistance to this form of enhancement.

Since vulnerability to smallpox and polio, or to measles, mumps, and rubella, is perfectly normal and natural (for those misguided enough to think that there is any virtue in what's normal or natural), then, if we alter human beings to affect their vulnerability to these things, we are enhancing them. We are interfering with perfectly normal and perfectly healthy human beings (babies or adults) to enhance them. Vaccines then are not "treatments," since the individuals vaccinated are not usually ill. They enhance precisely because they make changes to the normal physiology of humans which improve their resistance to disease and enhance their powers of survival. There may be some people who think this is wrong but you wouldn't want to be their child if you knew what was good for you!

Here again we see Rose's puzzling distinction between the mechanical and the chemical or between direct and indirect interferences with bodily chemistry.

Example 3: Genetic Enhancement

Is there something qualitatively different about intervening in the natural genetic lottery to improve upon (enhance) a naturally evolved or a naturally created genome? Many people clearly think so.[6]

A number of the world's leading laboratories are currently working on radical therapies which would also constitute enhancements. For example, David Baltimore's lab at Caltech is working on the possibility of engineering resistance or possibly immunity to HIV/AIDS and cancer into cells.[7] The benefits of this and related work around the world are incalculable. Whatever else they are, they would also constitute radical enhancement since, alas, immunity to HIV/AIDS or cancer is not part of "normal species functioning" or "species-typical functioning" for our species.

Baltimore notes that one way in which immunity might be engineered is by manipulating the immune system to resist cancers. "The immune system is genetically controlled," he says, "and it is possible to manipulate its mechanisms using gene therapy methods." Moreover, the way in which this might be done involves bringing "new genes into the cells using a virus as a carrier or vector." This work is a "grand challenge" and success is problematic, but the point for us, for society, is to decide whether we should hope Baltimore and others are successful or hope their work will fail.

It is tempting to speculate as to whether there are any people who think that Baltimore's work is wicked and should be stopped. I don't and I hope you don't, because if it were to succeed and to result in a genetic intervention that was safe and effective, millions of lives would be saved. Since it is difficult to imagine anyone hoping that Baltimore will be unsuccessful, we know that there are no principled objections to genetic enhancements per se nor to enhancements that prevent disease. From this we can conclude, anticipating more detailed arguments to come, that whatever is wrong (if anything) with enhancement, it is not that such modifications operate on genes or even on the germline, nor that they interfere with natural nor normal nor species-typical human functioning, nor that they reject some of the ways in which humans are constituted or find themselves with a "given" nature or set of capacities or vulnerabilities.

Francis Fukuyama is one among many who appeal to a nebulous and ultimately impenetrable notion to show why changes to human nature are absolutely unacceptable. He postulates a "factor *X*" that "entitles every member of the species to a higher moral status than the rest of the natural world."[8] And, trying to explain what he means by human dignity, he insists that "From a secular perspective, it would have to do with human nature: the species typical characteristics shared by all human beings qua human beings. That is ultimately what is at stake in the biotech revolution."[9] This is impenetrable because Fukuyama insists the factor *X* is what is left "when we strip all of a person's contingent and accidental characteristics away." He imagines we are left with "some essential human quality underneath."[10] Jonathan Glover[11] has identified the weaknesses of this view with unerring precision and I cannot here do justice to his critique. The essence of the problem with Fukuyama is that he gives no positive account of factor *X* which might persuade us either that it is worth preserving, or that if we lose it there will be hell to pay. Without this, Fukuyama is simply making a plea for precaution without any indication as to why precaution entails the preservation rather than the sacrifice of factor *X*. Indeed, if it could be shown that factor *X* could be enhanced, either to make it resistant to erosion or indeed in ways that boosted its essential *X*-ness, then presumably Fukuyama would have to endorse human enhancement.[12]

A more radical and even more intemperate objection to genetic enhancement of the sort which might be achieved if Baltimore's work bears fruit comes from George Annas.[13] In a diatribe against cloning and "attempts to cure or prevent genetic diseases and then to 'improve' or 'enhance' genetic characteristics," Annas seems to be against both the cure and the prevention of genetic diseases if it involves any changes to the human genome:

> Cloning, however, is only the beginning of the genetic engineering project. The next steps involve attempts to "cure" or "prevent" genetic diseases, and then to "improve" or "enhance" genetic characteristics to create the superhuman or posthuman.
>
> It is this project that creates the prospect of genetic genocide as its most likely conclusion. This is because, given the

> history of humankind, it is extremely unlikely that we will see the posthumans as equal in rights and dignity to us, or that they will see us as equals. Instead, it is most likely either that we will see them as a threat to us, and thus seek to imprison or simply kill them before they kill us. Alternatively, the posthuman will come to see us (the garden variety human) as an inferior subspecies without human rights to be enslaved or slaughtered preemptively.
>
> It is this potential for genocide based on genetic difference, that I have termed "genetic genocide," that makes species-altering genetic engineering a potential weapon of mass destruction, and makes the unaccountable genetic engineer a potential bioterrorist.[14]

Annas is aware that this will seem overblown (and says so in the very next line of his tirade). However, while to say this rhetoric is overblown is something of an understatement, it is problematic not because of all the rather strained huffing and puffing involved, but because, on the basis of mere speculation about future possible effects, Annas seeks to deny millions of people and eventually the entire population of the planet access to possible lifesaving and life-enhancing therapies.[15] To claim that "[i]t is this potential for genocide based on genetic difference, that I have termed 'genetic genocide,' that makes species-altering genetic engineering a potential weapon of mass destruction, and makes the unaccountable genetic engineer a potential bioterrorist" is about as plausible as saying that deliberately reproducing people of Jewish origin made all Jewish parents potential instigators of the Holocaust. Annas claims that the "genetic engineer is a potential bioterrorist" is based on the fact that in the first place the genetic engineer is unaccountable, and secondly she has brought into being an individual or individuals who are potential victims (or persecutors) because of a genetic difference that has been deliberately or knowingly created. This analogy is telling because all parents are unaccountable. Indeed, parents are more so than scientists, who in most countries have to have their work approved by ethics committees and, unlike parents, undergo lengthy training and education. Moreover, scientists are more extensively subject to national laws and regulations and international conventions. Moreover, where parents create children whom history has shown are likely to be different in a way that promotes hostility, discrimination, and sometimes murder, there is obviously a potential for genocide based on difference. With enhanced groups of people, just as with ethnic or religious or national groups of people, the therapy of choice is surely to operate on the mindless prejudice that creates the hostility rather than to stigmatize as terrorists or potential murderers, and indeed outlaw, those who deliberately create or perpetuate such identifiably "different" groupings.

I have in the past emphasized, contra Annas, that, far from constituting a threat to the preservation of the human genome, human reproductive cloning is the only method of reproduction that preserves the human genome intact (just as it was).[16] This is obvious, since sexual reproduction does not preserve the human genome but constantly varies it through an almost random combination of the genomes of the two parents. Cloning, in that it repeats a given genome, is the technique that can claim priority as method of genomic conservation. Equally, as Sarah Chan has aptly pointed out,[17] only universal cloning (by abolishing genetic difference altogether) can remove any temptation to, or potential for, "genocide" based on genetic difference. If Annas followed his own arguments, he would be the most staunch and fearless advocate for human reproductive cloning.

Annas, Kass, and other critics treat regenerative medicine, and cell-based therapies with enhancing properties, as if the enhancing properties were dispensable and disposable add-ons, as if regenerative medicine could be therapeutic without engaging the mechanisms that do the enhancing. But this is to misunderstand the way the science will work. We are unlikely to be able to separate the enhancing and therapeutic powers of drugs and techniques, as we noticed when considering Baltimore's work. This dual role of new therapies and techniques will be a recurring theme of this book, and their probable inseparability is an essential reminder of the complexity of the choices involved.

Double Effect

Of course the notorious doctrine of double effect could be invoked to suggest that it is permissible to use enhancements, as the primary intention or "first effect" is therapeutic and the second, enhancing, effect can be condoned because it is the inevitable but unwanted side effect of therapy. However, while such different levels of effect are distinguishable, they do not do the job required, which is to absolve the agent of responsibility for second effects. Anthony Kenny[18] pointed out long ago that if I get drunk tonight knowing full well, foreseeing, that I will have a hangover in the morning, it would be odd to say that I get drunk with the intention of having a hangover in the morning. However,[19] if for some reason having a hangover were morally significant or carried with it criminal responsibility, the fact that the foreseen hangover was unintended would not cut much moral ice. Suppose I am a pilot and if I fly with a hangover I may crash the plane, or I must testify in a crucial trial tomorrow and with a hangover I know I will forget the relevant evidence. While it is true, if you like, that I do not *intend* to have the hangover, it will not be true that I am innocent of the consequences of that hangover, either morally or (probably) criminally.[20]

Example 4: Chemical Enhancement

There are numerous and varied candidates for chemical cognition enhancers; a recent report listed literally scores[21] of possible candidates from vitamins to amphetamines, from herbs to brain–computer interfaces. Two of the most popular are

- methylphenidate (Ritalin), a cognition enhancer with a good evidence base, which is widely used for improving various aspects of cognition in children and adults,
- modafinil, which enhances wakefulness and alertness and is identified for possible enhancement of the functioning of pilots, long-distance drivers, and military personnel.

Steven Rose, talking of the possible use of so-called "smart" drugs like Ritalin, asks:

> Is it cheating to pass a competitive examination under the influence of such a drug? Polls conducted among youngsters make it clear that they do regard it as cheating, in the same way that the use of steroids by athletes is considered to be

> cheating. However, the military at least has no qualms about such enhancement. U.S. pilots in the recent Iraq war were said routinely to be using the attention-enhancing and sleep reducing drug modafinil (Provigil) on bombing missions.[22]

The cocktail of choice for military pilots is apparently amphetamines on the way out to hype them up for combat and modafinil on the way back to keep them awake and alert to get home safe.

The claims Rose is making here are complex and perhaps confused. If we ask why athletes using steroids are considered to be cheats, one obvious answer is that because the use of performance-enhancing drugs in competitive sport is banned their use must be clandestine, and it is cheating because it is contrary to the rules and an attempt to steal an unfair advantage. If, however, the rules permitted such use, then the advantage would not be "unfair" because it would be an advantage available to all.[23]

In the context of education we have noted that Rose and indeed many others think that

> [i]t is alright to buy educational privilege for one's children by paying for private tuition, but dubious to enhance their skills by feeding them drugs.

Now, buying educational privilege in a context in which not all can afford to do so is certainly unfair in some sense. But if we defend people's rights to do this it is because we see education as a good and we feel it is right to encourage people to provide goods for their children and wrong to deny them these goods even if not all can obtain them. Indeed, it is somewhat misleading to talk of buying educational privilege or advantage. It is possible that people may be seeking an edge, a relative advantage for their children, but it is more likely (and perhaps more decent) that they are simply seeking excellence, the best for their kids, rather than seeking an unfair advantage over others. Some say "when it comes to examinations, excellence just means being better, more successful" but this is not so. Examinations are not an end in themselves: they are supposed to measure excellence or at least ability in the subjects examined. They are the measure of excellence at, say, philosophy but do not constitute excellence in philosophy. Of course, to seek excellence, to do the best for your kids when others cannot match your efforts, will probably also confer an advantage, but is it doubtful ethics to deny a benefit to any until it can be delivered to all. The same is true of many other goods that cannot be equally provided for all. We will return to this point in a moment. It is not then wrong to attempt to do the best for your kids by providing them with goods that may, because they are unevenly distributed or differentially usable, also confer advantages on them relative to others.[24] Access to performance-enhancing drugs is obviously unlike education in that it is not considered to be an intrinsic good: "education for its own sake." However, it is clearly an instrumental good in the same way that a healthy diet is good, in that it conduces to health and longevity and the only remaining issue is the possible side effects of the drugs—the safety, in other words.

Positional Goods

Many people are on long-term drug treatment. I myself, like many men of my age, take daily aspirin and statins. These are widely available, but even if they were not and I could afford them or access them when others could not, I would still take them. I take these drugs not to get the better of my fellow men and women (as a so-called "positional good" to improve my position relative to others) but to give to myself the best chance (in absolute terms) of a long and healthy life. If they are available to all, I lose nothing and indeed I hope that they are or will become available to all. I take them not for advantage but for my own good. Fairness does not require that I should not try to protect myself because others cannot; it does not require that benefits should not be provided to any until they can be made available to all. Fairness might require that we make all reasonable attempts to achieve universal provision. True, taking aspirin is not straightforwardly (or, perhaps one should say, not obviously) a means to a positional good in the same way that taking a performance enhancer before an exam the results of which will mean that some will succeed while others fail would be. But of course if my taking aspirin while others do not means that I live and they do not, I seem to have a positional advantage (despite the fact that my competitors, since many of them will be dead, occupy no position at all). Just as it is not wrong to save some lives when all cannot be saved, it is not wrong to advantage some in ways that also confer a positional advantage when all cannot be bettered in those ways. However, it is of course wrong to save some but not all when all could be saved in order to advantage those who benefit. The discussion of the claim that advantaging more people by using any particular enhancement was or was not possible in any particular set of circumstances is complex and is way beyond the scope of the present discussion.

I favor and defend enhancements as absolute rather than as positional goods. I defend them because they are good for people not because they confer advantages on some but not on others. I am therefore uninterested in any collective action problems that result from their use for positional advantage rather than for the betterment of individuals or of humankind. The morally justifiable enhancements owe their moral justification to the fact that they make lives better, not to the fact that they make some lives better than others. Therefore, the collective action problem that results from the fact that people invest in enhancements—either to get an edge or to protect themselves from being made worse-off by others having them when I do not in a way that leaves everyone poorer and no one better off than anyone else—is a problem for them and not for me or for the enhancing technology. "Serves them right," you might say, and you'd be right! The ethical justification for or defense of enhancements is not that they do or might confer positional advantage but that they make lives better.

Consider a different example. We, in the United Kingdom, are trying (not very seriously in my view) to make kidney transplants available to all who need them, as it is believed the Belgians, the Spanish, and the Austrians have managed to do.[25] Even when that is achieved we know that thousands in the rest of the world cannot obtain the transplants they need. In India, for example, there are about a hundred thousand new cases of end-stage renal failure each year and only about three thousand transplants are performed.[26] We do not (and surely we should not) say that we will perform no more transplants here until the needs of all those in India can be met. And since we currently do not even have enough kidneys available for our own population, we do not think that fairness demands that we suspend our transplant program pending a sufficient supply for U.K. needs. Perhaps this is partially because in seeking to provide enough donor organs for our own population we are not trying to gain an unfair advantage over the people of India; we do this not selfishly but altruistically, knowing that the altruism will fall short of universality. It is not, however, partisan. If and when we have a surplus of donor organs for the United Kingdom, I hope and expect we will donate that surplus abroad, just as other countries that have achieved a surplus do.

Elsewhere[27] I have argued that there are no good moral reasons to prefer to help compatriots rather than strangers; but there are often powerful practical and political reasons why altruism and even obligation must sometimes begin (but not end) at home. We have space for just one obvious example: in the case of transplants the measures required need the backing of law and of a medical and regulatory system that protects donors and recipients.[28] This is not easy to provide internationally, although not of course impossible.

So when enhancements make life or lives better they are justified if they do just that if they also confer positional advantage that is no part of the justification and will in fact always constitute a moral disadvantage of their use, although whether this disadvantage constitutes a decisive argument against either the use or the permissibility of the enhancement will depend upon many other factors, among which are the degree of advantage, the degree of unfairness it creates, and the likelihood of the unfairness being minimized over time or by other factors such as compensation. Some of these factors will occupy us immediately or indeed later in this book. Others may be left unresolved either because they are simply insoluble or because they are beyond me!

Priorities and Distributive Justice

It is often suggested that enhancements (and also other high-tech and/or expensive procedures) should never be a priority until more basic treatments and welfare provision can be offered to all. There is much to be said for the sentiment, and for the moral consciousness expressed in this idea, but little for the evidence or arguments which might sustain it.

We have already considered that it is doubtful ethics to deny a benefit to some unless and until it can be provided for all. It is also doubtful economics and doubtful policy.

No one can be ignorant of the fact that procedures which start expensive, rare (even elitist), and risky often become widely available, if not universal, cheap, relatively safe (safe enough given the balance of risk and benefit), and widely accessible. In recent times, spectacles, portable timepieces (watches and clocks), radiophonic communications, radar, computers, access to the Internet and to satellite technology, mobile phones, and bicycles, not to mention motor transport, radio and television, and photography, film, and digital technology have followed this path (with mixed benefits to individuals and societies it must be admitted). In medicine, access to a physician, vaccines, antibiotics, transplants, contraception, and many drugs has become commonplace and generally available in industrialized (high-income) societies and widespread even in lower-income countries. There is of course reason to fear the escalating costs of high-tech medicine, but the point for the present argument is that products and procedures need to start somewhere if they are to get anywhere. This means that unless we permit and possibly fund the development, we never benefit from the product or procedure and wide (if not universal) access could never occur. There is no reason to believe that things will be different for human enhancement. If we banned innovations unless and until they could be made available to all, it is probable that they would never be (have been) developed. We cannot know this for sure but it is highly probable. More certain (almost as certain as almost-certainty permits) is that they would be considerably delayed with all the targeted and collateral damage that benefit delayed typically causes.

Of course, if the advantages of particular enhancements were significant or the costs of not receiving them sufficiently substantial, governments or private agencies might (should) step in to fill the gap.

There is no simple answer to the question of justice of access to enhancements any more than there is to justice and access to health care or other technologies.[29] This book is about the ethics of making available and accessing human enhancement, and about the moral and social impact of interventions in the natural lottery of life and in the course of evolution. How these are to be funded is another question entirely. We have examined and will continue to examine some of the constraints on access and funding. Very few new technologies have been given a universal priority. Antibiotics, developed as they were at the outset of World War II, were put into large-scale production by the United States following their success in treating septicemia, to which open war wounds were susceptible.[30] The near universality of smallpox vaccination has eradicated smallpox and the same (it is hoped) will soon be true of polio. These are rare examples of global provision. As indicated above, there is no moral case for delaying access to any treatment or technology with health benefits until we are in a position to provide equitable and universal access. The more beneficial the technology, whether it be therapeutic or enhancing, the greater the moral imperative for wide and equitable access. Some technologies (bicycles, motor transport, and mobile phones) have become widely available because they are highly attractive and useful, and others (vaccines, antibiotics) because national and international programs of access have been initiated. It is far too early to know how access will develop for the more radical and innovative forms of enhancement.

Example 5: Life Extension

Other groups[31] are working on life-extending therapies using a combination of stem cell research and other research into the ways that cells age to both regenerate aging or diseased tissue and switch off the aging process in cells. Again, to live several hundred years and perhaps eventually to become "immortal" is no part of normal species functioning for our species.

The ethics of trying to do such a thing are complex. I have discussed these complexities at length elsewhere and will do so in chapter 4.[32] For the moment we should reflect that none of them are powerful enough to derogate from an important truth. That truth is that lifesaving is just death-postponing with a positive spin! If it is right and good to postpone death for a short while, it is difficult to see how it would not be better and more moral to postpone death for longer—even indefinitely.

Regenerative Medicine

If dramatic increases in life expectancy are ever achieved, it is most probable that they will occur primarily through the use of regenerative medicine, harnessing the regenerative powers of stem cells to repair and replace diseased or damaged tissue. These same powers that can repair and replace diseased or damaged tissue may, in a healthy individual, augment normal functioning. That is why regenerative medicine may never be simply or merely therapeutic, but is likely always to have an enhancing dimension.

Consider the following, as yet imaginary, fragments of a dialogue between doctor and patient:

> I can successfully treat your heart disease, but there's a downside. I will be using regenerative medicine so, as a result of my treatment, your cardiovascular system will be healthier, so you will live longer; you will have enhanced life expectancy…

or

> I can treat the damage you have sustained to your brain but the problem is that the regenerative therapy I will use will regenerate your brain and the likely outcome is that not only will full function be restored, but in fact your brain will function much better than before, with enhanced memory and intelligence…. I am so sorry!

Who would think this an unacceptable side effect of treatment? Who would think that such people should be left untreated rather than allow these, by hypothesis, unacceptable consequences to occur?

I will return to the issue of increased life expectancy in chapter 4.

Risk

Of course, once Baltimore (or someone else) is successful, we will want to be sure that we are not tampering with healthy human beings in a way which will harm rather than benefit them or in which the risks are too high. But this insistence on rigorous risk assessment and on only proceeding if in all the circumstances of the case the risks are acceptable is a feature not only of all medical and scientific advance but of all human decision making whatsoever. Some have imagined that enhanced superhumans will be Frankenstein's monsters, powerful, malevolent, and uncontrollable. We should not of course deliberately create such beings or creatures where there is foreseeable danger of their becoming such beings. Insofar as such possibilities are unforeseeable they can no more be guarded against than the occurrence of any other possible but unforeseeable consequence of any type of intervention whatsoever.

But this is not a general or even a more acute problem with enhancement. It is a problem with any risky procedure, whether it be sex, drugs, or rock-n-roll, eating fatty foods, road transport, or vaccination and gene therapy.

Three further problems merit brief attention.

The Precautionary Principle[33]

UNESCO's International Bioethics Committee (IBC), reflecting on the ethics of tinkering with the genes, has maintained that "the human genome must be preserved as common heritage of humanity."[34]

A number of questionable assumptions are involved here. The first is that our present point in evolution is unambiguously good and not susceptible to improvement. Second, it is assumed that the course of evolution, if left alone, will continue to improve things for humankind or at least not make them worse. The incompatibility of these two assumptions is seldom noticed.

However, the common heritage of humanity is a result of evolutionary change. Unless we can compare the future progress of evolution uncontaminated by manipulation of the human genome with its progress influenced by any proposed genetic manipulations, we cannot know which would be best and hence where precaution lies.[35] Put more prosaically, it is unclear why a precautionary approach should apply only to proposed changes rather than to the status quo. In the absence of reliable predictive knowledge as to how dangerous leaving things alone may prove, we have no rational basis for a precautionary approach which prioritizes the status quo.

The fatuousness of the precautionary principle was exposed by the immortal F. M. Cornford in his seminal *Microcosmographia Academica* as "The Principle of the Dangerous Precedent":

> The Principle of the Dangerous Precedent is that you should not now do an admittedly right action for fear you, or your equally timid successors, should not have the courage to do

> right in some future case.... Every public action which is not customary, either is wrong, or, if it is right, is a dangerous precedent. It follows that nothing should ever be done for the first time.[36]

Playing God

Another commonly held objection to deliberate interventions in the human genome, or in evolution, is the idea that the human genome and Darwinian evolution are "natural phenomena or processes" and that there is some sort of natural priority of the natural over the artificial. This is often coupled with an argument from superstition: that it is tempting fate or divine wrath to play God and intervene in the natural order.

These suggestions are superstitious, fallacious, or, more usually, both. If it were wrong to interfere with nature we could not, among many other things, practice medicine. People naturally fall ill, are invaded by bacteria, parasites, viruses, or cancers and naturally die prematurely. Medicine can be described as "the comprehensive attempt to frustrate the course of nature."[37]

What is natural is morally inert and progress dependent. It was only natural for people to die of infected wounds before antibiotics were available or of smallpox and polio before effective vaccines.

Thomas Hobbes famously took a more realistic view of nature:

> [A]nd which is worst of all, continual fear, and danger of violent death; and the life of man, solitary, poor, nasty, brutish and short.[38]

Hobbes defended a social and political solution to the problems of the state of nature. If he had had available to him knowledge of the technologies we now have and are developing, he might well have opted for changing not simply the state of nature, but also the natural state of man.

If, as we have suggested, not only are enhancements obviously good for us, but that good can be obtained with safety, then not only should people be entitled to access those goods for themselves and those for whom they care, but they also clearly have moral reasons, perhaps amounting to an obligation, to do so.

3 | What Enhancements Are and Why They Matter

Enhancements are so obviously good for us that it is odd that the idea of enhancement has caused, and still occasions, so much suspicion, fear, and outright hostility. In particular, two influential sets of arguments by philosophers who do not in principle oppose enhancements have nonetheless attempted to place severe constraints on the legitimacy of using enhancement techniques. These objections have been made in terms of either the motivation or the objectives to be achieved on the one hand or on the tests that enhancements have to meet on the other, tests that constrain enhancements in ways that seem hard to justify.

We do not need a new or complex definition of the term "enhancement." In the context of interventions which impact on human functioning, an enhancement is clearly anything that makes a change, a difference for the better.[1] This is simply what is ordinarily meant by the term. We cannot necessarily say in advance what will constitute a change for the better, since we cannot predict all future states of the world or states of the individuals who might wish to consider an opportunity to make things better. What we can say, and what I will argue in detail in this chapter, is that enhancements are not plausibly defined relative to normalcy, to normal species functioning, nor to species-typical functioning; rather, as we shall see, a different account altogether must be given.

I will here consider three essays on enhancement, two of which seem to place unsustainable constraints on the use of enhancement technology as much as on the concept of enhancement itself. I will use these accounts as a way of trying to state more clearly what the moral and indeed political justifications for enhancements are and to further refine the arguments of chapter 1.

We will start with two important essays that discuss enhancement in some depth. In doing so I will look again, very briefly, at a now rather antique discussion of enhancement in my own book *Wonderwoman and Superman*.[2]

The essays, two of the most interesting on these themes, are by Norman Daniels[3] and Allen Buchanan and coworkers.[4] Much of what I want to say is critical of the approach taken in these two essays, but they constitute icebergs in the ocean of discussions of enhancement. They must be negotiated in one way or another if we are to proceed.

Wonderwoman and Superman was concerned with the ethics of both changing for the better the genetic constitution of the individual and changing people on a scale that would affect both society and the world at large, and with the question of what "change for the better" means. Society-wide changes are now often framed in terms of changes to populations. I argued that if the gains were important enough (sufficiently beneficial) and the risks acceptable, we would want to make the relevant alterations and be justified in so doing, indeed that we would have an obligation to make such changes. The ethical question seems still to be the same as then envisaged, and whether any proposed changes amount to changes in human nature, or to involve further evolution, seems ethically uninteresting. In particular, whether the enhancements might be judged to involve creating a new species, "a new breed," or amount to "self-evolution" or "post humanism" or "transhumanism" are not moral issues.

Beauty and the Beast

In *Wonderwoman and Superman*[5] I discussed modifications to humans that will be advantageous in terms of powers and capacities, and might change people in ways that made them oddities to nonenhanced individuals and perhaps initially even to one another. This is not the place to revisit these issues at great length. Suppose we were to make it possible for humans to "have eyes in the backs of their heads." This might be advantageous (although I imagine not greatly so, since we can look around with our present ocular configuration). However, this might also make people thus modified so sexually and arguably aesthetically repulsive that it is bad for you, all things considered.[6] While I maintain that the abnormality of a characteristic relative to other humans does not affect its value, if the abnormality was also odd in a way that attracted hostility or opprobrium to the enhanced individuals, the consequences of that oddity might make things worse overall. We would have to weigh up costs and benefits carefully in each case. If the benefits were great, the simple fact that the alteration made us either nonhuman or very odd would not be decisive. Suppose that further depletion of the ozone layer made humans very vulnerable to melanoma and it was discovered that green skin afforded complete protection. Suppose further that a safe intervention would change skin pigmentation to the required shade of green. I am sure I would go green, and if I had to make the intervention in newborns I would do it for my children. Others might prefer their children normal and cancerous. I would not impose on them, but I hope they would permit me to save my own life and that of my kids. My kids might have a hard time until all their friends were dead, but I imagine there would be enough caring greens like me to provide them with more durable companions—and they would have the last laugh.[7]

The "Transhumanist" Agenda

The use of the terms "transhumanist" or "transhumanism" is much in vogue, but these terms seem to imply an agenda. Espousal of such terms can seem to be a way of characterizing (and often embracing) a movement or quasi-religion which promotes, encourages, and indeed has as its objective the creation of a new species of "transhumans." This idea has, I believe, no special merit aside from the ways in which the changes that (might) lead to the creation of a new species are justified and indeed mandated by the good that they will do for us and our successors. To say you are a transhumanist is like saying you are a "born-again Christian" or a "fundamentalist Muslim." It is both a program and an identity. I have no transhumanist program or agenda. I do think there are powerful moral reasons for ensuring the safety of the people and for enhancing our capacities, our health, and thence our lives. If the consequence of this is that we become transhumans, there is nothing wrong with that, but becoming transhumans is not the agenda; improving life, health, life-expectancy, and so on is, however, not only part of a defensible moral agenda, it is a mandatory dimension of any moral program.

The Humanist Agenda

We noted in the previous chapter that UNESCO's IBC, reflecting on the ethics of tinkering with the genes, has maintained that "the human genome must be preserved as common heritage of humanity."[8] Very many people talk as though being human was a moral imperative. The use of expressions like "what it is to be human" or "she is a person of exceptional humanity" or "it is only common humanity to act in this way" or "humanity requires that we do this" seem to reinforce this idea. The contrast with being human is of course intended to be "being less than human," being "subhuman" or "inhuman" or, worse, being "bestial" or "like an animal." Here the extent to which expressions like "human" or "inhuman" are used act as indices of morality and perhaps also culture. Hitherto exceptional conduct or ability which seems to be positive has been regarded as evidence of what humanity is capable of or as a benchmark of all that is best about being human. Some exceptional abilities or some capacities for exceptional conduct may in the future also involve something which we are forced to regard as involving a step change beyond humanity or a change of degree which also involves a change in kind, from humankind to some other kind. If we are feeling euphoric, we may be inclined to say that such changes "transcend" common humanity and maybe also uncommon humanity. Changes to the human genome may be thought of themselves to involve a move away from our current species identity; more radical changes may make such a reevaluation of our nature inevitable. In chapter 4 we will consider one such change, increased longevity, which may eventually amount to the creation of immortals. Since for millennia humans have used the term "mortals" or its equivalent to distinguish our kind from other kinds—"immortals," for example, who may also be gods—it may be difficult to avoid this change in our understanding of ourselves and what is fundamental to our nature.

It is difficult, for me at least, to see any powerful principled reasons to remain human if we can create creatures, or evolve into creatures, fundamentally "better" than ourselves. It is salutary to remember that we humans are the products of an evolutionary process that has fundamentally changed "our" nature. We have noted the absurdity of our common ape ancestors in Africa getting together with a simian agenda to block evolution so that simian nature would be preserved as "the common heritage of simian kind." If that had happened you would not be reading this book nor would I have written it! (No bad thing you might think, and you might be right; but then you would not be thinking anything because you would never have existed.)

In recent literature a distinction has often been drawn between the supposed moral significance of changes to the germ line and changes to the somatic line; that is, between changes that involve the gametes, and so are transmissible via reproduction to future generations, and those which, because they do not affect the gametes, remain one-off changes to the individual him- or herself. Changes to the germ line, once demonstrated to be safe enough, would surely be the therapies of choice owing to their massive advantage in terms of efficiency. If the change is important enough to make in the individual, then, if it can be made on the germ line and passed on indefinitely to future generations, that simply avoids the necessity of a separate alteration to each and every future generation. However, we should note that radical enhancements which raise issues of species identity or membership, if introduced into the germ line, might create new types of beings and not simply mutant variants of an existing type.

The Moral Motivation for Enhancement

Daniels has reminded us[9] that the high safety standards routinely now applied to human subject research will often rule out enhancements, whereas Buchanan et al. have suggested that the motive for all enhancement therapy must be, or at least include, the pursuit of equal opportunities. If the ethics of changing human nature or changing the nature of individuals (creating freaks, as Daniels calls this) is principally a matter of cost–benefit analysis, safety versus advantage, then the question as to what an appropriate safety standard might look like becomes crucial. This is one of Daniels's main themes so we will start with Daniels's stimulating and fruitful analysis of what might be meant by changing human nature.

Daniels and the Fruit Fly

> To summarize the lessons from my excursion into fruit fly nature: fruit fly nature is a *population* concept: to characterize the nature of fruit flies we must aggregate phenotypic variations across allelic variations among fruit flies. It is a *dispositional* concept, since the phenotypic traits we take to be the "nature" of fruit flies vary within some range under different conditions. Finally it is a *selective*, theory-laden concept: not every trait of fruit flies is likely to be considered part of their nature, but only those that we use to explain something of importance to us about them. Putting these points together, the concept of fruit fly nature applies to what we consider the central explanatory features of the *phenotype* fruit flies manifest under some range of conditions....
>
> So we can modify human nature, but it takes a very tall tale. We must affect the (or at least a) whole population of humans, and we must do so with a trait central to that nature.
>
> Still, what the modification does at most is change the particular individual's nature into something beyond human.... By itself, this does not alter human nature. It creates freaks. If it operated on a population level, we might well, as we have seen, count it as a change in human nature. A world full of mind readers instead of liars would be one in which we would not encounter human nature as we know it. Not many traits fit this description, I suspect.[10]

Daniels believes the concept of human nature is problematic in ways that can be elucidated by reference to the fruit fly.

Daniels then is rightly skeptical about the idea that most changes about which people worry can in a meaningful sense be thought of as changes to "human nature." He thinks, surely correctly, that for most likely changes this is the wrong question to ask or a misleading specification of the task. He concedes that we can change the nature of some humans (if not human nature), creating the equivalent of freaks. However, while we may have justifiable motives for running the inherent risks in order to ameliorate dysfunction, Daniels believes we do not have such justification for improving upon an "otherwise normal trait":

> For example, if we are trying to ameliorate or eliminate a serious genetic disease, or disease for which there may be some genetic or other medical remedy, the probability of potential benefit from the experimental intervention may plausibly outweigh the certainty of catastrophic illness. But if we are trying to improve on an otherwise normal trait, the risks of a bad outcome, even if small, outweigh the acceptable outcome of normality. So we cannot ethically get there from here.

I believe this argument has great force.

Talking of a cognitive trait, improvement in short-term memory loss, he says:

> We would have to know that the increased short-term memory involved here actually plays a role in enhancing the more complex cognitive task rather than ... interfering with it.... Without some clear sense of these complex issues, we could not have any confidence that improving the component capability has the intended or desired effect on the more complex one. And all of this information goes well beyond the standard worry that the intervention itself carries with it risks that noninterference lacks. In short, a careful human research protocol would most likely stop this experiment in its tracks.

Daniels notes that in some fields (he cites plastic surgery) surgeons routinely experiment by modifying techniques "on the hoof" but he seems to feel that

> our incomplete adherence to appropriate patient-protection concerns should not count as an argument in favor of ignoring them in the case of medical—genetic or not—interventions to improve on human cognitive or behavioral traits.

It is difficult to know whether the caution that Daniels regards as mandatory is better founded than that of UNESCO already considered. It is also unclear whether Daniels's precautionary approach is intended to apply only to cognitive and behavioral traits or whether he would also include enhancements to achieve longevity, or resistance to disease. These may all involve the same degree of "incomplete adherence to patient-protection concerns" and "the standard worry that the intervention itself carries with it risks that noninterference lacks." With the result that "a careful human research protocol would most likely stop" experiments like these in their tracks.

These arguments illuminate the ethics of enhancement in important ways. In *Wonderwoman and Superman*[11] I considered many of these issues in some detail and concluded on an optimistic note; Daniels has not provided arguments powerful enough to destroy this optimism:

> For my own part I welcome the possibility of a new breed of persons with life chances not available to us now. Of course engineering genetic protections of various kinds into the human genome will not automatically make the world a better place to live in, nor will it necessarily make people happier. We will still have to work as hard as ever to reduce disease, new diseases are after all always liable to arise. We will still have to work as hard as ever to reduce prejudice, including prejudice against the new breed, to combat injustice, to eliminate poverty, starvation, cruelty and the thousand unnatural shocks that flesh is heir to; as well as the natural ones.
>
> But the fact that we cannot cure everything has never been an argument for failing to cure something, particularly when it is something that causes pain, misery and premature death.[12]

Curing Dysfunction and Enhancing Function

It is traditional to draw a distinction between repairing or curing dysfunction on the one hand and enhancing function on the other. Those who, like Boorse,[13] Daniels,[14] and others, define disease in terms of a departure from normal species functioning or species-typical functioning are often drawn to this distinction because for them enhancement is also a departure from normal species functioning and species-typical functioning. Repairing dysfunction is on this view restoring species-typical functioning from below, so to speak, whereas enhancement is departing from species-typical functioning in an upward trajectory. On this view, treating disease is restoring species-typical functioning, whereas enhancement is interrupting species-typical functioning. However, it is of course always true that restoring species-typical functioning is enhancing for the individual concerned unless they are functioning above par and the restoration is injurious or damaging to them. Most of what passes for therapy is an enhancement for the individual relative to her state prior to therapy.

Another problem for this distinction between therapy and enhancement arises because the very same interventions which, for some, might help restore normal, or as I would prefer to characterize it "better or more adequate" functioning, may for others be radically enhancing. Regenerative stem cell treatments for people with brain damage would clearly be therapy. The same treatment in those with normal or undamaged brains might, however, enhance brain function; likewise, treatment for memory loss in damaged brains might enhance memory in normal brains.

Elsewhere I have criticized the therapy–enhancement distinction, relying, as above, on the friendly ozone layer to provide the telling illustration:

> Suppose due to further depletions to the ozone layer, all white skinned people were very vulnerable to skin cancers on even slight exposure to the sun, but brown and black skinned people were immune. We might then regard whites as suffering substantial disabilities relative to their darker skinned fellows. And if skin pigmentation could be easily altered, failure to make the alterations would be disabling....

> [I]n such circumstances whites might have disabilities relative to blacks even though their functioning was quite species-typical or normal.[15]

We do not die of old age but of the diseases of old age.[16] It is species typical of us to die of these as we normally do, but it is not necessarily necessary that we do. If we could systematically treat these diseases in a way that enabled tissue to regenerate (using stem cell therapy perhaps) and at the same time switch off the aging process in cells, this too would be enhancing but would be another case in which treating disease in particular ways also constituted enhancement. Systematically treating the diseases of old age with the result that people lived substantially longer to the extent perhaps that some would even become immortal would appear to constitute both therapy and enhancement. This, however, is only because treating disease seems typical of therapy, not because normal species functioning does or can play any role at all in the argument.

However, as Boorse and Daniels define disease, treating the diseases of old age would not be therapeutic (indeed, the diseases of old age would not be diseases) in any sense because diseases of old age are species typical (or of course constitute normal species functioning), and it is species typical and a part of normal functioning that we cease to function in old age and that we die.

In an earlier essay on this theme Daniels claimed that

> it is our norms and values that define what counts as disease, not merely biologically based characteristics of persons.... Pointing to the line between treatment and enhancement is not, then, pointing to a biologically drawn line but is an indirect way of pointing to the valuations we make.[17]

If this is true, we must ask what the relevant evaluation is. The answer must surely be that we have a strong rational preference not to be harmed in a particular way and it is this that leads us to make the evaluation that draws the line. Occam's razor, that wonderfully enhancing surgical tool, comes to our aid here and shows us that we do not need to use our values to create a spurious distinction (the normal–abnormal divide), which Daniels insists is the route to distinguishing therapy from enhancement. After the surgical strike we are left not with a process that leads from our norms, through a definition of disease and the normalcy–enhancement distinction to a moral difference between therapy and enhancement, but one which moves directly from our values to a rejection of harm and an acceptance of benefit whether called therapy or enhancement. This shows that the enhancement–therapy distinction does not depend on conceptions of normalcy at all and the moral imperatives either to provide therapy or enhancements come from the fact that we value minimizing harm. Normalcy plays no part in the definition of harm and therefore no part in the way the distinction between therapy and enhancement is drawn.

Of course in view of these examples those who wished to cling to the therapy–enhancement distinction could simply say "yes, therapy and enhancement sometimes amount to the same thing, but not always." But now consider that many routine therapies—vaccination, for example—are enhancement technologies for the simple and sufficient reason that they enhance and the moral significance of the distinction as well as its utility collapses. It does not draw either a morally significant or an explanatorily significant distinction and so fails utterly to be useful. We will shortly return to the question of what is a morally relevant distinction in this field.

This forces a reconsideration of the distinction between therapy and enhancement and a reconsideration of the question of our motives for and the justification of our interference in the natural lottery of life. Here it is fruitful to look at the attempts by Buchanan et al. to grapple with this.

Our Commitment to Intervene in the Natural Lottery of Life

In one of the most famous and influential philosophical books on this subject, Allen Buchanan, Dan Brock, Norman Daniels, and Dan Wikler have argued that the motive we have for intervening to treat or cure disability (and the same would presumably go for the treatment of disease) is "for the sake of equal opportunity." Buchanan et al.[18] say "some of our most basic social institutions reflect a commitment to intervening in the natural lottery for the sake of equal opportunity." To be sure, the main concern of Buchanan et al. is to show how the concept of equality of opportunity requires extension to embrace what Scanlon and others have termed a "brute bad luck" conception of equal opportunity—a conception which believes in intervening to mitigate disadvantaging factors which are beyond the control of the subject of those factors.[19] This is a constantly repeated, but not universal, gloss on the moral reasons Buchanan et al. give for therapeutic or even enhancing interventions. Again,

> [i]n other words, equal opportunity has to do with ensuring fair competition for those who are able to compete *and* with preventing or curing disease that hinders people from developing the abilities that would allow them to compete.[20]

Buchanan et al. are self-consciously following Daniels and Rawls here: "As Norman Daniels has argued, the case for a moral right to health care relies, at least in part, on the fact that health care promotes equal opportunity."[21] Here the move from simply extending a conception of equality of opportunity to confirming that conception as an important part of the moral right to health care becomes clearer. It is this idea—the idea that the moral reasons we have for pursuing health or for enhancing the functioning of human beings (in short, our "commitment to intervening in the natural lottery" of life) has much to do with equal opportunity or the "ability to compete"—that is genuinely bizarre.

Of course, equality of opportunity is something we should try to maximize in the delivery of improvements in health or in functioning, but it seems only tenuously and contingently connected with our moral reasons for so doing. Suppose there was a painful condition which affected some people but not others. It did not, however, affect people's ability to compete or affect the pursuit of opportunities. I believe we would have exactly the same compassionate motive and moral reasons for intervening in the natural lottery to remove this condition as we would have if it was also "competition affecting." This latter factor seems to add nothing to our moral reasons for alleviating the condition.

The reverse is also true. Where a feature frustrates equal opportunity, that fact also provides a sufficient and complete moral justification for trying to remove it. What it does not do is provide a necessary justification.

The commitment to intervene in the so-called "natural lottery" would (or should) surely be compelling quite independently of any contribution it makes to equal opportunity, although, as I have indicated, in pursuing health and/or enhancement for the good reasons we have for so doing, we should attempt to ensure equal opportunity to access such goods. But the claim that "the case for a moral right to health care relies, at least in part, on the fact that health care promotes equal opportunity" seems false. Equal opportunity might in some circumstances be a further additional reason to support the moral right to health care, or of course a separate one, but to regard equal opportunity as an essential part of the moral basis for such a right or indeed the moral motivation for establishing such a right is doubtful. I have argued elsewhere[22] that where the provision of health care will prevent harm to human beings, the moral argument for delivering that care is as complete as it needs to be. Equal opportunity can say something about selecting who to help in conditions of scarcity but it seems an inappropriate candidate for a factor which might have any priority in explaining or justifying "a commitment to intervening in the natural lottery."

Suppose, now, that all people were affected by a debilitating condition so that there was no *inequality* of opportunity, but, inter alia, the opportunities for all were reduced. The unnecessarily reduced opportunity would itself constitute sufficient moral reason for "intervening in the natural lottery," not for the sake of equal opportunity (nor surely for the sake of competition), but for the sake of enhanced opportunity or enhanced functioning. Equally, if, as we discussed when considering enhancement, a new protective treatment were to be developed, then, even though before implementing it all would be equally disadvantaged by not having the treatment, the moral imperative to introduce it would not refer to equal opportunity at all and that case would not be enhanced by any such appeal.

Buchanan et al.[23] note that

> [i]t is possible, however, that some natural inequalities are not departures from normal species functioning[24] but nonetheless so seriously limit an individual's opportunities that he or she is precluded from reaching the threshold of normal competition. In such cases, genetic intervention

might be required if it were necessary to remove this barrier to opportunity.

Anticipating the question as to the precise circumstances in which this might be true, Buchanan et al. answer: "Whether it does will depend on what the normal distribution of various characteristics is and how that relates to the most fundamental requirements for successful participation in social cooperation in a given society."[25] It seems implausible to think that either normal species functioning or successful social cooperation are the key ideas that license interference in the natural lottery of life, and it is not hard to see why.

Consider again the enhancement technologies we discussed earlier: those involving engineering resistance to HIV and cancer into cells or enhancing the life expectancy of human beings. These two sets of possibilities would be radically enhancing to the extent that if we manage to find ways to make such changes permanent, to insert them into the germ line, for example, "we" might no longer be human; we would perhaps have evolved into a new distinct species. Whatever one thinks of these prospects, it is parochial in the extreme to imagine that our ethical response to them would depend on whether or not the failure to introduce these possibilities would "so seriously limit an individual's opportunities that he or she is precluded from reaching the threshold of normal competition." We have to ask what is the motive to introduce these and other new therapies in the first place? It cannot be to restore normal species functioning because immunity to cancer is not part of this. Nor would we plausibly want to make people immune to cancer in order to help people to reach "the threshold of normal competition" because we have a much better reason for so doing as I demonstrate below. Normal competition and normal species functioning do not constitute reasons for considering the introduction of these and many other new therapies or enhancements. Nor could the idea of equal opportunity help us here: opportunity perhaps, but not equal opportunity.[26]

Imagine twin sisters, both of whom suffer from cancer. One is curable, the other not. We do not think that equal opportunity requires us to leave the curable twin untreated because we cannot treat both. And our moral motive and reason for treating the curable twin has nothing to do with equality. It has everything to do with saving a life that can be saved or alleviating pain, suffering, and distress when presented with an opportunity so to do.

The moral imperative for David Baltimore's work, for example, is that it is required for what Hobbes referred to as "the safety of the people"[27] and for what others have called "beneficence or nonmaleficence," or welfare. These are imperatives quite independent of their impact on or compatibility with equal opportunities or equal justice. A moment's reflection shows why this must be so. There is a vast shortfall in the availability of donor organs for transplantation in the world, certainly in the United States and in the United Kingdom. As I have already shown, while we cannot treat all who need lifesaving transplants, we treat as many as we can and we do so because to fail to do so would cost lives. We do not say we will perform no transplants at all unless and until we can secure equal access to transplants for all those who need them. Hopefully we allocate access to those organs that become available in ways that are consistent with equal opportunity; but the reason why we save lives in this way is not to secure equal opportunity or to secure access to "normal competition."

What is clear is that the moral motive for using technology to intervene in the natural lottery of life is for the sake of the goods that this will bring about. Equality of opportunity may sometimes be one of these goods, fair equality of opportunity may also be sometimes one of these goods, it may even be fairer than (presumably unfair or less fair) equality of opportunity, but neither are plausible candidates for a prime moral imperative in these contexts. More usually, equality of opportunity, fair or less fair, will be a constraint on the way the goods may be legitimately be achieved. Saving lives, or what is the same thing, postponing death, removing or preventing disability or disease, or enhancing human functioning are the more obvious and usually the more pressing reasons. These are the primary reasons we have to prevent or mitigate disability and to treat or cure disease.

Daniels and the Limits to Enhancement

In view of the moral imperatives that underlie enhancement just identified, Daniels's precautionary approach fails in its prime protective function. Consider again his argument for precaution:

> For example, if we are trying to ameliorate or eliminate a serious genetic disease, or disease for which there may be some genetic or other medical remedy, the probability of potential benefit from the experimental intervention, may plausibly outweigh the certainty of catastrophic illness. But if we are trying to improve on an otherwise normal trait, the risks of a bad outcome, even if small, outweigh the acceptable outcome of normality. So we cannot ethically get there from here.
>
> I believe this argument has great force.[28]

Daniels's argument here has two essential elements. In the first he suggests that the benefits of intervening in the natural lottery may outweigh the risks where there is a "certainty of catastrophic illness" as the alternative. It is important here to be clear as to whether the certainty of illness is statistical or personal. For example, where we seek to engineer immunity to HIV/AIDS into the genes of future people there is no certainty that any individual will be exposed to HIV, let alone contract it, but the risk to individuals and to populations is such that the lack of certainty in any particular case seems irrelevant to the justification of taking the requisite risk. Of course, the degree and magnitude of risks that might be worth taking to protect ourselves from major diseases are importantly relevant, but the assessment of these seems not to depend upon the certainty of contracting the disease in question. The same was true of vaccination for smallpox, polio, and other diseases. This is a case where there may be no direct benefit to the subject of the research or therapeutic procedure but where the procedures are justified by their expected benefits to society or to populations and to the interests of the subject more broadly conceived.[29]

This of course challenges some of the standard elements of the international conventions that purport to govern research. For example, when Daniels notes above that "a careful human research protocol would most likely stop this experiment in its tracks," he is perhaps thinking of paragraph 5 of The Declaration of Helsinki, which notes: "In medical research on human subjects, considerations related to the well-being of the human subject should take precedence over the interests of science and society."[30] However, the coherence of this provision is challengeable and depends crucially on how narrowly the interests of the subject are conceived.[31]

The second element of Daniels's argument places great weight on the idea that if we are "trying to improve upon an otherwise normal trait the risks of a bad outcome, even if small, outweigh the acceptable outcome of normality." It is not clear here what Daniels thinks the link is between the normality of a trait and the acceptability of the outcome. He seems to imply that normal traits are of themselves acceptable, perhaps because he thinks that disease is always a departure from the normal and hence that normality is a disease-free state. But, as we have seen, disease and unnecessary death are hideously normal and although they impair normal functioning at some stages of life they do not do so in old age, when it is normal to succumb to them. In old age the diseases of old age are part of normal functioning; what is not normal in old age is to be disease free, have perfect vision and hearing and no increased vulnerability to colds, flu, and other diseases.

Clearly, if we are trying to improve upon an otherwise acceptable trait, then small risks may not be worth running. The problem even with this formulation is that awareness of the possibility of improvement is likely to affect the acceptability of a trait. Shortsightedness at a certain age is normal and was perhaps acceptable before spectacles; and the pains of childbirth were thought not only acceptable but mandatory even after the advent of effective analgesia; painful birth is certainly perfectly normal species functioning.

Increases in life expectancy are an even clearer case. It is certainly not normal for humans to live beyond 100 years and so far as we know no one has yet lived beyond about 122 years, although many believe it probable that the first 150-year-old person is already alive and some serious commentators believe the first 1,000-year-old person is already alive.[32] Dying of old age is acceptable[33] only because it is regarded as inevitable. Absent the inevitability, the acceptability is problematic. We can, perhaps somewhat artificially, try to separate dying of old age and dying of the diseases of old age. We have noted that dying of the diseases of old age is normal, but perhaps if we were able to combine regenerative treatments of the diseases of old age with the ability to switch off the aging process in cells, even longer and healthier life expectancy might be achieved. This would even more clearly have little to do with normal functioning and yet this fact seems of little help in thinking about either the ethics or the desirability of such enhancements, let alone about the question of which risks it might be worth running to achieve them.[34] If we are trying to improve on an otherwise normal trait, like old age, it seems implausible simply to assume or to stipulate that "the risks of a bad outcome, even if small, outweigh the acceptable outcome of normality."

The normality of the trait in question is clearly doing no work at all in the assessment of its moral acceptability or of the risks it might be worth running to change things. If we subtract the acceptability from the normality we are left with nothing of moral, nor of argumentative force. Traits in short are not acceptable (in the normative rather than of course the simply descriptive sense of "acceptable") because they are normal, they are acceptable because they are worth having. If they are not worth having, or if they are worth not having, their normality seems bereft of interest or force. This, it seems to me, obviously applies to saving life, that is to postponing death or to increases in longevity, resistance to the diseases of old age and to those which strike at any time like HIV, cancer, and heart disease. Whether it also applies to enhancements in cognitive function like memory or other processing skills, or to personality, is more problematic only insofar as the benefits are more problematic in many ways. There seems to be no difference in principle here and thus no difference in the relevant ethical considerations.

When Daniels says "we cannot ethically get there from here" this seems doubtful; and I am unclear as to why he believes "this argument has great force." The problem seems to be an unjustified assumption that normal traits are acceptable by reason of their normality and that the risks of new "treatments" are justifiable only when the alternative is an inevitable catastrophic disease. Weaken or qualify the argument so that the inevitability applies on a population but not an individual basis, and so that normal traits are only acceptable when they are desired or beneficial, and the argument revives, but only at the cost of abandoning the therapy–enhancement distinction altogether, and only if disease is defined relative not to normal species functioning or species-typical functioning but to *possible functioning*.[35] The only remaining question then is whether or not the possibilities are worth attempting given the benefits on offer and the degree or magnitude of the risks involved. Reference to normality or to disease is not required in order to answer this question.

What is required is, as we have noted, the recognition that the moral imperative and the most usual moral motive for using technology to intervene in the natural lottery of life is for the sake of the harms this will prevent and the goods that this will bring about. Saving lives or—what is the same thing—postponing death, removing or preventing disability or disease, or enhancing human functioning are examples of these goods. The only remaining question is which risks are worth running to achieve these goods either for individuals or for populations. Where the risk is only to the individual, she surely should decide for herself. Where the risk is to future individuals or to future populations, society must have a say, which is why it is societies rather than populations which should stay in our central focus. The principles upon which societies may decide these things are not a subject to be tackled in this chapter.[36] What can now be said, however, is that societies have a responsibility to address these issues and that their responsibility is not clarified or simplified by reference to what is normal, natural, or species typical. Nor is it much use thinking about how disease or illness or disability have traditionally been defined, nor is reference to equal opportunity except helpful insofar as this refers to equal opportunities to access the relevant technologies.

The moral question is and remains: how beneficial will be the proposed enhancements and whether or not the risks of achieving them are worth running for individuals or societies? The moral imperative is the safety of the people and the duty to compare risks with benefits, not on the basis of the normality of the risks or of the benefits, or on the basis of their contribution to equality of opportunity, but on the basis of their magnitude and probability, on whether they will save life or—what is the same thing—postpone death and on how much harm and suffering may be prevented or avoided by the enhancements in question.

Dandelion Technology

Rehearsing his commitment to defining treatment as "anything to do with maintaining normal function," Daniels, with a somewhat magisterial waft, records that he sees "the issue as largely semantic and not central to the distinction between treatment and enhancement. Indeed, had we no vaccine for HIV/AIDS, and no genetic modification providing protection, but instead found a diet that included dandelion leaves gave protection, we would hardly say we had enhanced human nature."[37] I do not see the differences between Daniels's approach and my own as "largely semantic" and I do not see how Daniels can dismiss them in this way either, since whether something is treatment rather than enhancement makes crucial differences in the moral reasons we have to make interventions both in Daniels's system and in that of Buchanan et al.

Daniels's analogy with dandelion leaves seems to me to be unconvincing. At first sight it looks as though it would be like saying in the context of a debate about the ethics of radical surgery: "indeed had we no option of bilateral mastectomy for breast cancer, but instead found a diet that included dandelion leaves gave protection, we would hardly say we had undertaken radical surgery." The point of Daniels's example is to counter a definition of enhancement, but this it signally fails to do. Of course, I was not assuming that all methods of achieving an objective would be enhancing if any of them were, it is obviously more likely that only some (but not all) methods of protection constitute enhancement, but even here the distinction is difficult to draw and more difficult to sustain; ultimately it collapses altogether. Consider Daniels's dandelions: depending on an understanding of their mode of operation and the durability or permanence of the protection afforded, we might come to think of dandelion leaves as the ultimate enhancement technology! When Daniels says, "[i]n another sense, providing the vaccine stimulates a normal immune system to produce antibodies it is capable of producing so that individuals and a population are protected against serious diseases,"[38] he begs the question: because the immune system would not normally do this, doing it is therefore not part of normal functioning.

On the issue of mode of operation Daniels says:

> I have generally characterized vaccines as part of the array of "treatments" for departures from normal functioning—although this is stretching the idea of treatment to include preventive measures, a stretch I have made explicit in my

writings. In effect, anything to do with maintaining normal function falls under the scope of "treatment" as opposed to enhancement.

But this simply assumes the point at issue, which is whether making a normal immune system do something it normally doesn't do constitutes "normal functioning." We can imagine a radical enhancement, say a stem cell treatment that caused brain cells to proliferate, increasing brain capacity and function by a factor of ten. Daniels might still say "this treatment stimulates a normal brain to produce cells it is capable of producing so that individuals and a population have bigger brains." This sleight of hand enables Daniels, if he wants, to define literally anything as therapy so long as it operates on a normal physiology and causes it to do something abnormal. If it successfully makes a normal bodily system do something, then obviously this is something that it is capable of doing. On this view the elixir of life, operating on a normal physiology and making it immortal, would not in Daniels's view be an enhancement!

Many of those hostile to enhancement trade, as Daniels does, on insisting that radical enhancement of functioning, so that people cease to be vulnerable to diseases or conditions which it is normal or species typical of us to succumb to, does not count as enhancement at all. Perhaps they think of enhancements in terms of things like "bat sonar" that would enable people to avoid bumping into objects in the dark or breeding red-nosed reindeer called "Rudolph" or human equivalents, so that sleighdrivers delivering Christmas presents would not bump into chimney pots. Such objections have in mind that enhancements should be, ought to be,[39] things that change the nature of the human condition. They think perhaps that "immunity to cancer just allows normal species functioning—it does not augment it," conveniently forgetting that cancer is a major killer precisely because it is all too normal a part of our species functioning. If we could eradicate cancer and heart disease, that would certainly change the nature of the human condition.

Therapy and Enhancement Are Not Mutually Exclusive

Daniels and others treat treatment and enhancement as if they are necessarily mutually exclusive: an intervention is either one thing or the other and different consequences follow from how interventions are classified. His assumption that vaccines are treatments and indeed are also preventive measures is of course unassailable. But my point is that treatments or preventive measures which protect humans from things to which they are normally vulnerable or which prevent harm to that individual by operating on the organism, by affecting the way the organism functions, are also necessarily also enhancements. That goes for drugs, foods, vaccines, implants, and whatever else, whether they are "natural" or "artificial" and regardless of whether they are part of a self-selected diet or treatment or are recommended or prescribed as a result of scientific research. The boundaries between treatment and enhancement, between therapy and enhancement, are not precise and often nonexistent, nor are these categories mutually exclusive as Daniels seems to think they must be. If we wear a suit of armour and the slings and arrows of outrageous fortune simply bounce off, that may not constitute an "enhancement" of the human being (though it seems to me that even here the protection–enhancement distinction, in the way that Daniels sees it at least, is on the borderline of fracture to the point at which I myself am unclear as to which side of the border we are on). We already use prostheses and implants (heart pacemakers, for example) as technological and mechanical aids or enhancements. Although the more familiar prostheses replace, say, lost limbs or teeth, or even breasts, they often do things that the originals could not do! In the future, silicon chips and other electrical and computer prosthetic devices or implants may assist and enhance memory, vision, and many other functions. Nanotechnology may enable us to make some of these so tiny and lightweight as to be unnoticeable. On the other hand, if we change human physiology or metabolism or genetics, or whatever, by dietary modification, vaccination, stem cell therapy, dandelions, or other organic means so that the organism becomes immune or resistant to slings and arrows, or automatically repairs the damage they do from within, the changes may seem more natural and more intrinsic to the human organism as traditionally understood. Either way, the distinction between therapy and enhancement, between protection and improvement, cannot be coherently or consistently maintained. Whatever else this distinction does, it fails to identify two mutually exclusive categories, and those wonderful dandelions will have enhanced human functioning.

Another reason the issue is not largely semantic is because consequences flow from whether or not something is classified as or "semantically constitutes" normal functioning. As Daniels (2007) admits: "anything to do with maintaining normal function falls under the scope of 'treatment' as opposed to enhancement"—for Daniels maybe! However, for many people, including most doctors and those who run health care systems, treatment constitutes an obligation in a way that enhancement (as yet) does not.

The overwhelming moral imperative for both therapy and enhancement is to prevent harm and confer benefit. Bathed in that moral light, it is unimportant whether the protection or benefit conferred is classified as enhancement or improvement, protection or therapy. For Daniels it does matter, because for him "anything to do with maintaining normal function falls under the scope of 'treatment' *as opposed to* enhancement" (my italics), and for many purposes (although for none of mine) therapy is given a higher priority than enhancement. For Buchanan et al. the most important moral imperative seems to be equality of opportunity. I have shown that both these views fail to capture the moral purposes which both therapy and enhancement standardly and most appropriately serve. I have also shown that existing and accepted enhancements, such as vaccination, make more sense when seen as conferring benefits or protecting from harms than when seen as serving either the interests of equal opportunity or those of restoring normal functioning.

Since the therapy–enhancement distinction goes nowhere in identifying where on the permissibility–impermissibility distinction an intervention lies, we need to move to the main question which is: are there limits to freely chosen enhancements, and if not would there be legitimate constraints on opportunities offered to enhanced individuals—in terms of education, employment, freedom to compete in sports, arts, and other commercial and recreational activities? I have tried to indicate some answers to this question in my 1992 book *Wonderwoman and Superman*, but for present purposes the answer may be summarized as indicated above, namely the overwhelming moral imperative for both therapy and enhancement is to prevent harm and confer benefit. The remaining questions concern the detail of how to calculate where that balance lies in any particular case.

4 | Immortality

The Holy Grail of enhancement is immortality.[1] Increased longevity and its logical extension, some would say its reductio ad absurdum, immortality, have a long history. The human imagination is familiar with the idea of immortals and mortals living alongside one another and doing all sorts of things together, including of course having sex and fighting! The *Iliad*, the *Odyssey*, the Bible, the Qur'an, the Ramayama, and Shakespeare's plays have all made such ideas familiar; and even modern classics have taken seriously the possibility of immortality. In his celebrated five-part trilogy *The Hitchhiker's Guide to the Galaxy*, Douglas Adams imagines a man who had achieved immortality by accident:

> To begin with it was fun, he had a ball, living dangerously, taking risks, cleaning up on high-yield long-term investments, and just generally outliving the hell out of everybody.
>
> In the end it was the Sunday afternoons that he couldn't cope with, and that terrible listlessness which starts to set in at about 2.55, when you know that you have had all the baths you can usefully have that day, that however hard you stare at any given paragraph in the newspapers you will never actually read it ... and that as you stare at the clock the hands will move relentlessly on to four o'clock, and you will enter the long dark teatime of the soul.[2]

Despite the apparent pessimism of this passage many people would be prepared to endure "the long dark teatime of the soul" in exchange for immortality.[3] Indeed, there is much evidence that suggests that many people are willing to trade off quality of life for longevity.[4] From the pact of Faust, celebrated by writers from Marlowe to Goethe, Berlioz, Thomas Mann, and Mikhail Bulgakov to Bram Stoker and his vampires,[5] to choices made by cancer patients with a terminal diagnosis,[6] the evidence is strong that people want extra lifetime even at substantial costs in terms of pain, quality of life, and even when outcomes are highly uncertain.

Life-extending therapies and optimistic discussions of their promise and probable effect are an increasing dimension of serious scientific and philosophical discussion.[7] If such therapies ever become reality,[8] and if our bodies could repair damage due to disease and aging "from within,"[9] the effects not only on personal health and survival but also on society and on our conceptions of ourselves and of the sorts of creatures we are would be profound.[10] If we could switch off the aging process,[11] we could then, in Lee Silver's words, "write immortality into the genes of the human race."[12]

I am not of course suggesting that we are on the verge of discovering the elixir of life, although, as noted in chapter 3, there is at least one living person who has studied these matters and who thinks that the first 1,000-year-old human is already alive. I am, however, interested in exploring the ethics of attempting to prolong life and even to create immortals if that should ever prove possible.

Immortality Is Not Invulnerability

We should note that immortality is not the same as invulnerability, and even "immortals" could die or be killed. Accidents, infectious diseases, wars, and domestic violence would all take their toll; and although we might hope for progress in combating existing diseases, the development of new threats, as HIV/AIDS, resurgent avian flu, and the emergence of Creutzfeld–Jakob disease have demonstrated, may increase rather than reduce human vulnerability over time. If we add to this the diminishing effect of proven therapies such as antibiotics through the emergence of resistant strains of bacteria, it is difficult to predict the likely levels of "premature" deaths in a future in which increased life expectancy is developing and spreading through the human population. This is one reason why predictions about the connection between increasing life expectancy and overpopulation are often reckless.

Life Extension Saves Lives

When we save a life, by whatever means, we simply postpone death. Since lifesaving is just death-postponing with a positive spin, it follows that life-extending therapies are, and must always be, lifesaving therapies and must share whatever priority lifesaving has in our morality and in our social values. So long as the life is of acceptable quality (acceptable to the person whose life it is)[13] we have a powerful—and many would claim overriding—moral imperative to save the life, because to fail to do so when we can would make us responsible for the resulting death.[14]

Five main sorts of philosophical or ethical objections have been leveled at life extension. The following claims have been made.

(i) Life extension would be unjust.

(ii) It would be pointless and ultimately unwanted because of the inevitable boredom of indefinite life.

(iii) It would in any event be nugatory or self-defeating because personal identity could not survive long periods of extended existence. I may wish to be immortal but in the end it wouldn't be "me," so the project fails.

(iv) It would lead to overpopulation and perhaps the end of reproduction.

(v) Finally, it is claimed that life extension would be prohibitively expensive in terms of increased health care costs.

We will be examining each of these objections in due course.

Global Justice

One thing we do know is that the technology required to produce dramatic increases in life expectancy will be expensive. For existing people, with multiple interventions probably required, the costs will be substantial. To make modifications to the embryo or even to the gametes prior to conception, people will have to be determinedly circumspect about procreation and will probably need to use reproductive technologies to have their immortal children. Even in technologically advanced countries therefore, "immortality" or increased life expectancy is likely to be confined to a minority of the population. In global terms, the divide between high-income and low-income countries will be increased, with low-income countries effectively denied access to the technology that might make some of their citizens immortal. The issue of the citizens of rich countries gaining further advantages over the poor will rightly disturb many. How are we to understand the demands of justice here?

Parallel Populations

A feature of life-extending treatments which has seldom been thought through is the fact that as treatments become available we will face the prospect of parallel populations, of "mortals" and of "immortals" existing alongside one another.[15] Thus, the problems of global justice will be repeated in those societies able to implement life-extending therapies. Just as there will exist parallel societies, some able to provide immortalizing therapies and some not, so, within those societies that have the technology and the resources required, there would exist parallel populations of mortals and immortals. This of course is precisely the destiny for which the poetic imagination has prepared us, literally from "time immemorial."

While the creation of such parallel populations seems inherently undesirable and even unfair, it is not clear that we could, or even that we should, do anything about such a prospect for reasons of justice. If immortality or increased life expectancy is a good, it is doubtful ethics to deny palpable goods to some people because we cannot provide them for all. And this unfairness is not simply contingent, a function of a regrettable, but in principle removable, lack of resources.

There will always be circumstances in which we cannot prevent harm or do good to everyone, but no one surely thinks that this affords us a reason to decline to prevent harm to anyone in particular. As I pointed out earlier, if twins suffer from cancer and one is incurable and the other not, we do not conclude that we should not treat the curable cancer because this would in some sense be unjust to the incurable twin. We don't refuse kidney transplants to some patients unless and until we can provide them for all with renal failure. We do, however, have a clear ethical responsibility to maximize benefits and in cases of scarcity to ensure that the question of which of those who could benefit receives the treatments should be decided according to some morally defensible principle of distribution.

The introduction of any new complex and/or expensive technology raises these problems. The impact on global justice or on justice within societies is important and must be addressed; it is a principled objection, but not an objection in principle to the introduction of life-extending therapies. The principle requires that strenuous and realistic efforts be made to provide the benefits of the technology justly and as widely as possible, not that the benefits be denied because of the impossibility of ensuring adequate justice of provision.

We should also note that we live in a world, and indeed in most cases also in a country, that already has such parallel populations. We know, for example, that people in the southeast of the United Kingdom live longer than those in the northwest and that people in industrial cities have shorter life expectancy than those in leafy suburbs. Life expectancy in 2001–2003 for males in Manchester (where I live) was eight years less than that for a fashionable part of London, Kensington and Chelsea (from a life expectancy of 71.8 years in Manchester to 79.8 in Kensington and Chelsea).[16] On the world scale it is even more dramatic, the A–Z of life expectancy extending from Andorra at the top with 83.5 in 2002 to Zambia at the bottom with 37.2.[17]

Speaking personally of our terrible situation in Manchester, eight years' difference may not sound a lot, but in stepwise improvement in treatment for advanced solid tumors in colorectal cancer, for example, increases in survival of between five and twelve months are regarded as dramatic improvements.

As someone who wants to live forever and who happens to live in central Manchester, I feel this injustice keenly! However, the solution, as with transplants, is surely not to level down nor yet to put improvements in some places on hold until they can be provided for all.

Regenerative Medicine

Remember that immortality is not unconnected with preventing or curing a whole range of serious diseases. It is one thing to ask the question, should we make people immortal? and answer in the negative, and quite another to ask whether we should make people immune to particular diseases or to treat heart disease, cancer, dementia, and many other diseases and decide that we shouldn't, because a "side effect" of the treatment would be an increase in life expectancy. We are, then, unlikely ever to face the question, should we make people immortal—yes or no? We may rather be called upon to decide whether we should treat a particular disease when we know an effective treatment will extend life span.

It might then be appropriate to think of immortality as the side effect of treating or preventing a whole range of diseases. Could we really say to people, "you must die at the age of thirty or forty or fifty, because the only way we can cure you is to do something which will also extend your life span"? Faced with such a choice an individual might well say, "let me have my three score and ten and then kill me if you must." Whether, given the quite pervasive and irrational hostility to euthanasia, societies would be willing to allow such bargains to be made is doubtful.

Longevity Is a Rational Good

Given that people want life and fear death it is difficult not to see longevity, and perhaps immortality, as a palpable good. Many have taken issue with this claim on two main grounds: either that indefinite life would eventually become terminally boring, or that over long periods of survival personal identity could not be maintained and so the survival of a particular individual would prove illusory. Elsewhere[18] I have criticized, and I believe decisively refuted, both these objections.

Suffice it to say that only the terminally boring are in danger of being terminally bored, and perhaps they do not deserve indefinite life. Those who are bored can, thanks to their vulnerability, opt out at any time. But those of us who do not have terminal failure of the imagination should be left to create new ways of enjoying life and doing good.

Personal Identity

Vertical Identity Change

Hans Jonas and Walter Glannon have suggested[19] that over the very long life of a Methuselah personal identity must fail, giving place to successive selves, and that it follows that prudential or self-interested motivation for continued existence must also therefore fail. However that may be, and for my own part I am utterly unconvinced that personal identity would fail in the circumstances suggested, it is easy to see that personal identity is not required for a coherent desire for indefinite survival. Suppose "Methuselah" has three identities, A, B, and C, descending vertically into the future and that C can remember nothing of A's life. But suppose the following is also true: A will want to be B, who will remember being A; B will want to become C, who will remember being B but possibly not remember being A. It is not irrational for A to want to be B and not irrational for A to want to be B partly because he or she knows that B will be able to look forward to being C, even though by the time she is C she won't remember being A. Thus, even if personal identity in some strict sense fails over time because of loss in continuity of memory, it is not clear that a sufficiently powerful motivation for physical longevity fails with personal identity. This would remain true however many selves "Methuselah" turns out to be. For myself I am skeptical as to whether failure of memory is nearly enough to cause failure of personal identity but here is not the place to argue the point.

A B C

B

C

Horizontal Identity Change

The point remains good for what we might term "horizontal" identity failure, although again I am skeptical as to whether cognitive enhancements or indeed other sorts of enhancement that might impact on personality or life history would lead to failure of personal identity.

Suppose chemical cognitive enhancers either involved dramatic personality changes as a side effect of their enhancing properties or that the enhancements were so marked that someone with such dramatic improvements in memory, intelligence, or concentration would predictably live a very different life to their former self to the extent that, in either case, the question of whether or not they would remain the same person was a real issue. Again the question of whether or not B or C, after successive enhancements or the time and buildup of decisions and choices that only the cognitively enhanced person would have made, would be enough like the unenhanced A to count as the same person would not destroy the rationality of A's enterprise. Here again A would have decided to create C, a different but closely related individual, and would have enough continuity with C to want to create C and to welcome and embrace the required personality changes that might raise issues of personal identity.

Interestingly, if Jonas, Glannon, and others are right and extreme longevity does raise issues of successive selves, successive different persons inhabiting the same body so to speak, then two prominent objections to longevity and other identity altering enhancements fail. For if the long-lived person is in fact a succession of only averagely old individuals, then claims that making people live longer is giving them an unfair advantage over those who do not benefit from life extension also fails. After all, there is no one who has this advantage, just a succession of different people who have long but not excessively long lives. Moreover, it is hardly harmful to any of them, let alone anyone else, that one full life gives way to another using the same body. Indeed, from the perspective of resources and overpopulation this multiplication of people without any multiplication of resources used by the extra numbers must surely be advantageous, as will the fact that these extra persons add not at all to the total population of the world, presumably have a good start in life, and don't require long periods of expensive education and equally costly dependence on parents, etc. Here, however, the successive new persons inhabiting the same body are no more responsible for any increases in population than are the successive new persons born in different bodies from those of their "parents."

Again for the horizontal case, even if the cognitive enhancement of A involves her eventual disappearance to be replaced by B or C, this would at worst involve the creation of extra individuals without the necessity of creating the extra bodies and the extra resources those bodies would consume or the extra opportunities or space those extra bodies would require.

Indeed, multiplication of selves in the same body is so ecologically sound, environmentally friendly, and population efficient that it might well become the preferred method of procreation for all except the most unregenerated eco-wastrels or sex-obsessed chauvinists.

Immortals Are Inhuman

Prominent among recent denigrators of the idea of life extension has been Leon Kass, who identifies the core question as the following: is it really true that longer life for individuals is an unqualified good? Kass has many arguments against life extension, all of which fail disastrously.[20] We have time to consider only his main objection:

> For to argue that human life would be better without death is, I submit, to argue that human life would be better being something other than human.... The new immortals, in the decisive sense, would not be like us at all. If this is true, a human choice for bodily immortality would suffer from the deep confusion of choosing to have some great good only on condition of turning into someone else.[21]

Insofar as this claim of Kass's relies on claims about psychological continuity over time, it has the problems we have already considered. However, Kass's argument seems to be suggesting a more simple objection: that since the (current) essence of being human is to be mortal, immortals would necessarily be a different type of being and therefore have a different identity. There is a sense in which this is true, but not I think in any sense in which it would be irrational to want to change identity to the specified extent. Someone who had been profoundly disabled from birth (blind, say, or crippled) and for whom a cure became available in his or her mid-forties would become in a sense a different person. They would lead a different type of life in many decisive ways. It does not follow that the blind or crippled individual has no rational motive to be cured. It would be both odd and cruel to say to them, as Kass presumably would have us do, "it is deeply confused to want to cease to be disabled because then you will no longer exist." We will return to the theme of disability in chapter 6.[22]

Population Policy

Many people addressing the question of life extension have assumed that such a possibility will have a disastrous effect on the world's population with the present generation living indefinitely and a procession of subsequent generations adding to the congestion.[23] However, this is by no means either a likely or even the most likely scenario. We have already noted some advantages to longevity and personality change in terms of fears of overpopulation. Even without these somewhat fanciful advantages, the effect of life extension on population will be a function of a number of different factors, the outcomes of which are all difficult to predict. The first is the degree of uptake, which itself will be heavily dependent on cost and availability of the therapies. Granting, as we have, that life-extending therapies will gradually become available, cost, risk, and uncertainty will mean that for a very long time the numbers of people availing themselves of such therapies will be a tiny proportion of the world's population. We have already noted a possibly increasing human vulnerability due to new infectious diseases or antibiotic resistant strains of bacteria. Again it is difficult to predict the continuing effect of these on population or how the advent of some immortals would affect the equation. Disease may well continue to be an effective leveler, improving its own technology as we improve ours. And of course immortal but vulnerable people will continue to die in accidents and from injuries received.

Steve Austad[24] has calculated the death rate of immortals in the United States due to such factors. Austad bases his calculation on the death rate of twelve-year-olds. Kids of twelve have stopped dying of childhood diseases and have not yet started dying of the diseases of old age. If you work out their death rate, you can extrapolate a probable death rate for immortals. Austad originally suggested this might indicate an average life-span of 1,200 years. He now writes:

> [H]ealth has improved in the U.S. since then. I recalculated this for the 2002 census data, and if we were now preserved in the health of a 12-year-old, life expectancy

would be 4,986 years. Not bad, living from the age of the early Pharaohs to the age of Mars landings.[25]

The End of Reproduction

Should we assume the necessity for, or desirability of, the creation of future generations? Is there a moral difference between a future that will contain *x* billion people succeeded by another *x* billion different people and so on indefinitely, or *x* billion people living indefinitely and replacing themselves on the (rare?) occasions when they are killed? Although, as we have noted, this is an unlikely scenario, posing the question in this stark form enables us to ask an important question: whether or not what matters morally is that life years of reasonable quality exist or that different people with lives of reasonable quality exist. Put in this way the problem assumes a familiar form—should we maximize life years or individuals' lives?[26] From the life-years perspective it ought not to matter how many new people the world will contain but simply how many life years of acceptable quality it will contain. Those who, like me, find the life-years approach unsatisfactory will be inclined to think that individual lives matter. But even so, it could consistently be held that it is the individual lives of existing people that matter, not how many new individual lives there will be.

However, the argument for making sure that there will be new generations is not settled by the outcome of the debate between those who think that future lives count equally with existing lives and those who do not. One group of such reasons has to do with the desire to procreate and the pleasures of having and rearing children.[27]

The second set of reasons has to do with the advantages of fresh people, fresh ideas, and the possibility of continued human development. If these reasons are powerful, and I believe they are, and if the generational turnover proved too slow for regeneration of youth and ideas and for the satisfactions of parenting, we might face a future in which the fairest and the most ethical course might seem to be to contemplate a sort of "generational cleansing."[28] This would involve deciding collectively how long it is reasonable for people to live in each generation and trying to ensure that as many as possible live healthy lives of that length. We would then have to ensure that, having lived a "fair innings," they died at the appropriate time in order to make way for future generations. Achieving this result by voluntary or ethical means might be difficult; attitudes to suicide and euthanasia might change, but probably not overnight. Although it might appear to offer a solution, I believe no nonconsensual form of generational cleansing would be defensible.[29] However, since the numbers of immortals would remain small for the foreseeable future, this is a problem that may never arise.

Immortality Is Cost-Effective

Søren Holm[30] has suggested that immortality, far from increasing health costs per individual, might actually dramatically reduce them; there might in short be an economic discounting argument for the public funding of "immortality" interventions. We should note that, although immortality or at least longevity is expected to eliminate many of the diseases of old age, there will still be terminal conditions to treat (at least for those who do die) and hence always some postponed health care costs for the dying.

If we assume that

- for both mortals and immortals there is the same period of old age with increased health care costs (say ten years, but the length does not matter for the argument) and the same costs of treatment during those years (let's say £10,000 on average),
- the mortals will reach this period in 70 years and the immortals in 1,000 years,
- there is a 1% per year rate of real economic growth,

then the present-day discounted costs of treating a person in 70 years' time will be £4,948, whereas the present-day cost of treating the same person in 1,000 years' time will be 43 pence! It thus makes economic good sense to invest now and postpone health care costs from 70 years into the future to 1,000 years into the future and, as is evident from the figures, it makes sense even if immortals were to have a much longer and more costly old age (because of the discounting, even a ten times increase in costs would not matter).[31] Add to this the probability that a greater number of immortals would die as the result of accidents rather than long drawn-out illnesses and the economic arguments grow stronger still.

Conclusion

For the first time in human history we face the prospect of a truly open future, involving perhaps infinite sequential as well as simultaneous opportunities, and stretching, open-ended, before the individual in a an unprecedented but truly liberating pathway. We should be slow to reject cures for terrible diseases even if the price we have to pay for those cures is increasing life expectancy and even the creation of immortals. Better surely to accompany the scientific race to achieve immortality with commensurate work in ethics and social policy to ensure that we know how to cope with the transition to parallel populations of mortals and immortals as envisaged in literature and mythology.

5 | Reproductive Choice and the Democratic Presumption

There are many arguments from many sides which purport to give reasons for limiting access to reproductive technologies and procedures which may facilitate the enhancement of individuals or permit illness, impairment, or disability to be removed or minimized. The issue is whether or not these arguments point to dangers or harms of sufficient seriousness or sufficient probability or proximity to justify the limitation on human freedom that they require.

The Democratic Presumption[1]

One of the presumptions of liberal democracies is that the freedom of citizens should not be interfered with unless good and sufficient justification can be produced for so doing. The presumption is that citizens should be free to make their own choices in the light of their own values, whether or not these choices and values are acceptable to the majority. Only serious real and present danger either to other citizens or to society is sufficient to rebut this presumption. If anything less than this high standard is accepted, liberty is dead.

This presumption is sometimes expressed as saying that citizens should enjoy the maximum liberty which is compatible with a like liberty for all. This way of putting the liberal presumption acknowledges that one legitimate limitation of the liberty of the individual is where its exercise limits the liberty of others, or threatens others with significant harm.

The alternative to a presumption of liberty is what John Stuart Mill[2] called the "tyranny of the majority." To avoid this tyranny, the presumption in favor of liberty can only be rebutted by showing that the exercise of liberty for some either infringes the like liberty for others, or causes real and present dangers of significant harm either to individuals or to society. It is not enough that others are made uncomfortable by its exercise, nor that they do not like it, nor that they find it repellent.

Upholding liberty, safeguarding a free society, is not cost free. One of the costs is that citizens must be prepared to accept that others must be free to do things that they themselves would not do, would not wish to do, and even things that make them uncomfortable or which they find repugnant. The liberty to do only those things of which the majority approve is no liberty at all.

Freedom of religion and freedom of conscience are good litmus tests here. Since for all monotheistic religions there can be only one true god, it follows that all nonbelievers are not only mistaken but heretical. To protect their souls, or to protect the one true faith, or to avoid offense to God, infidels should be suppressed. That we do not burn heretics is part of our commitment to freedom and to democratic values. Equally, if I judge you to be seriously morally wrong I must think you should not do what is wrong. But unless I can also show that what you propose to do, or are doing, is harmful to others or society, then a commitment to liberal democratic values means that I must leave you room to differ from me. What consenting adults do in private, and particularly what they do to themselves, is (almost always) their own affair. The exceptions must establish serious harm to others or society.

The burden of proof is not on those who defend liberty but on those who would deny it.

In most democracies (though not all) there is a presumption in favor of liberty. As Joel Feinberg put it:

> Whenever a legislator is faced with a choice between imposing a legal duty on citizens or leaving them at liberty, other things being equal, he should leave individuals free to make

> their own choices. Liberty should be the norm; coercion always needs some special justification.
>
> It is legitimate for the state to prohibit conduct that causes serious private harm, or the unreasonable risk of such harm, or harm to important public institutions and practices. In short, state interference with a citizen's behavior tends to be morally justified when it is reasonably necessary (that is, when there are reasonable grounds for taking it to be necessary as well as effective) to prevent harm or the unreasonable risk of harm to parties other than the person interfered with.[3]

Such a presumption means that the burden of justifying their actions falls on those who would deny liberty, not on those who would exercise it. If this is right, the presumption must be in favor of the liberty to access reproductive technologies and other means of founding families unless good and sufficient reasons can be shown against so doing.

Thus, those who would exercise reproductive liberty do not have to show what good it would do; rather, those who would curtail freedom have to show not simply that it is unpopular, or undesirable, or undesired, but that it is seriously harmful to others or to society and that these harms are real and present, not future and speculative, for if they were not, the presumption in favor of liberty would be at risk whenever imaginative tyrants could postulate possible, but highly unlikely, future harms.

It should go without saying that a right to reproductive liberty is not an entitlement to the cooperation of anyone in the exercise of that right, in the sense that if I have a right of reproductive liberty that does not mean that you (my desired sexual partner) or you (my physician) or you (my fertility expert) must necessarily cooperate with me in my pursuit of a family. Rather, it means that neither the state nor professional organizations, nor advisory or regulatory bodies may legitimately prevent any of you from willingly cooperating with me if that is what you choose to do. In short, it means that such cooperation must be neither illegal nor prevented in other ways, for example, by prohibitively costly taxes or prohibitively onerous administrative procedures.

Reproductive Liberty as a Basic Human Right

But more than being a simple exercise of liberty or personal preference, reproductive liberty has a serious claim to be a dimension of a fundamental human right. If it involved the exercise of a bare preference, like drinking coffee or playing tennis, its free exercise would still be a fundamental entitlement. However, if it can be shown to involve more than an assertion of an entitlement to exercise freedom of choice, but that it also involves the claim to exercise a choice that is part of a fundamental or basic human right, then the arguments against its exercise must be proportionately stronger and the harms that are claimed to result from its exercise must be proportionately greater. I believe that reproductive liberty has a good claim to be part of such a fundamental right.

As Ronald Dworkin has persuasively argued, making use of the distinctions between liberty as license and liberty as dignity or liberty as a bare freedom and liberty as a basic right:

> If freedom to choose [a good] is simply something that we all want, like air conditioning or lobsters, then we are not entitled to hang on to these freedoms in the face of what we concede to be the rights of others to an equal share of respect and resources. But if we can say, not simply that we want these freedoms, but that we are ourselves entitled to them, then we have established at least a basis for demanding a compromise.[4]

Dworkin has defined reproductive liberty or procreative autonomy as "a right to control their own role in procreation unless the state has a compelling reason for denying them that control."[5]

Julian Savulescu gives a classic twist to arguments about reproductive liberty, suggesting that the core idea derives from an element in John Stuart Mill's defense of liberty, which highlights the crucial role played by the freedom to experiment. Savulescu sets the idea out thus:

> Reproduction should be about having children who have the best prospects. But to discover what are the best prospects we must give individual couples the freedom to act on their own value judgement of what constitutes a life of prospect.

> "Experiments in reproduction" are as important as "experiments in living" as long as they don't harm the children who are produced. For this reason, reproductive freedom is important. It is easy to grant people the freedom to do what is agreeable to us; freedom is important only when it is the freedom for people to do what is disagreeable to others.[6]

I have argued that the key idea of reproductive liberty is surely respect for autonomy and for the values which underlie the importance attached to procreation.[7] These values see procreation and founding a family as involving the freedom to choose one's own lifestyle and express, through actions as well as through words, the deeply held beliefs and the morality which families share and seek to pass on to future generations.

Reproductive Freedom Embraces New Technology and Enhancement

Implicit in my defense of reproductive liberty and, I believe, in that of others such as Dworkin and Savulescu is the idea that such liberty must extend to the use of technology and methods of reproduction not envisaged by Adam and Eve. The legal theorist Antje Pedain has made this implication explicit, and elaborates the argument that I have failed to provide:

> Harris insists that human rights extend to the use of new technologies which expand our powers, options and ability to affect another person's fate and condition in ways and by means which were previously unknown.
>
> While it initially appears that this assumption might require some argument, on reflection this is not so. Technical advances often change the way in which we exercise our rights and freedoms, and thus broaden the practical scope of these rights. Freedom of movement now extends to moving around by car or plane, and not just by foot, boat or bike, and may tomorrow encompass flying to Mars in a rocket. Freedom of speech now extends to the distribution of newspapers, television, and Internet chatrooms, and not just to speaking at public assemblies and the like. If fundamental

rights and freedoms were not capable of protecting new ways of exercising them, their scope of application would shrink over time in that, with the advances of technology, the right in question would only cover some, instead of all instances of exercising the protected activity. The only way to prevent such a gradual erosion of fundamental rights and freedoms is to expand their range of application along with the changes of technology.

> Consequently, Harris is right in saying that the presumption of liberty applies not only to procreative techniques that achieve outcomes which, if circumstances were different, sexual procreation could achieve, but also to the use of techniques which may eventually enable prospective parents to achieve outcomes beyond what is possible by "natural" sexual procreation (for instance, the exchange of single defective genes for nondefective ones in embryos, and the creation of embryos by cloning or by combining genes from more than two existing human beings). Any restriction of this liberty requires a sufficiently weighty reason, and it is in this context that the important aspects in which the new forms of exercising a fundamental right differ from the ones which were previously known can and must be taken into account.[8]

Now of course freedom of movement did not extend to moving about by car *before* cars were invented. However, the scope and limits of basic freedoms, like freedom of speech and of movement, are not settled by the means of movement or the modes of expression current at particular moments in history. Rather, they can only be established by reflection upon the point and purpose of protecting such rights.

Ronald Dworkin insists that

> [t]he right of procreative autonomy follows from any competent interpretation of the due process clause and of the Supreme Court's past decisions applying it.... The First Amendment prohibits government from establishing any religion, and it guarantees all citizens free exercise of their

> own religion. The Fourteenth Amendment, which incorporates the First Amendment, imposes the same prohibition and same responsibility on states. These provisions also guarantee the right of procreative autonomy.[9]

The point is that the sorts of freedoms which freedom of religion guarantees—freedom to choose one's own way of life and live according to one's most deeply held beliefs—are also at the heart of procreative choices. Similarly, we must ask what freedom of movement or freedom of speech as fundamental rights are designed to do. It is not that moving and speaking are by themselves fundamental (although they are of course important), rather that they are essential for the exercise of the "freedom to choose one's own way of life and live according to one's most deeply held beliefs"; they are in short essential for a moral and political life.[10]

And Dworkin concludes that

> no one may be prevented from influencing the shared moral environment, through his own private choices, tastes, opinions, and example, just because these tastes or opinions disgust those who have the power to shut him up or lock him up.[11]

Thus, it may be that we should be prepared to accept both some degree of offense and some social disadvantages as a price we should be willing to pay in order to protect freedom of choice in matters of procreation.[12]

The nub of the argument is that

> [t]he right of procreative autonomy has an important place ... in Western political culture more generally. The most important feature of that culture is a belief in individual human dignity: that people have the moral right—and the moral responsibility—to confront the most fundamental questions about the meaning and value of their own lives for themselves, answering to their own consciences and convictions.... The principle of procreative autonomy, in a broad sense, is embedded in any genuinely democratic culture.[13]

Freedom and Enhancement

Thus, the freedom to access reproductive technologies and indeed enhancing technologies or procedures is at the very least protected by the democratic presumption. Where these procedures are part of reproductive decision making they may well be a dimension of a fundamental of basic human right already established and widely recognized.

Either way, the burden of proof is not on those who would exercise this liberty or right to enhancement to show what good it does; rather, it is on those who would limit it to show how and to what extent its denial is necessary to protect either the exercise of a like liberty for all or is required to protect others or society from real and present harms or dangers. If, as I have argued, we also have a positive moral obligation to enhance others if we can do so without compromising our own lives (or those of third parties or society) unduly or disproportionately, then again the freedom to do what we ought to do is one that any just and ethical society should grant.

Whether or not this freedom should be publicly funded is of course a separate question. What is clear is that it should not be curtailed by law, regulation, or intimidation.

The Obligations We Have to Future Generations

Everything we do today which has long-term consequences (which is almost everything we do) in some ways affects future generations.[14] There are many specific sorts of obligations we might think of in connection with future generations: not to spoil the environment for them or use up valuable natural resources or leave behind sources of danger like spent nuclear fuel, unexploded nuclear weapons, or other dangerous pollutants, and obligations concerning their genetic inheritance. All of these are part of our general obligation not to harm those who come after us.

There are two standard ways of bringing about harm. One is by using what have been called "positive actions" and deliberately changing things so that harm results. The other equally effective way is by deliberately leaving things as they are, knowing that harm will result—using so-called "negative actions."[15] The idea that there is no significant moral difference between acts and omissions with the same consequences is sometimes called the "acts and omissions doctrine." Although once controversial, this view now seems to me incontrovertible.[16] It would obviously be as bad, say, to pollute the environment so that the incidence of cancer increased by twenty percent as it would be not to remove a naturally occurring environmental hazard which we could easily remove and the failure to remove which would cause the same degree of harm. Or, it would be as bad for a mother to fail to take a drug during pregnancy that would remove a disabling condition from her child as it would be to deliberately take a drug that would bring about the disability.[17]

Our obligation not to cause harm to future generations has the same positive and negative faces. We must not act positively so as to cause harm to those who come after us, but we must also not fail to remove dangers which, if left in place, will cause harm to future people. Thought of in this light there is a clear dilemma about enhancement. On the one hand, we must not contemplate enhancements which will adversely affect our descendants. On the other hand, we must not fail to make changes that could be made which will avoid harm to future people or which would benefit them in ways that cannot be achieved unless these enhancements are put in place.

We must in short weigh up the probability of harm occurring as a result of what we do against the probability of harm occurring if we fail to take steps now to prevent its future occurrence. In some cases the dilemma may be acute; we may simply not know enough to be able to make reasonable judgments as to the various probabilities involved. In such cases we should err on the safe side. It is a singular fact that the safe side is always supposed to be the side of inaction or preserving the status quo, although this belief is entirely without foundation.

It is clear that we should not fail to remove a danger which is real, present, and highly probable for fear that there may be some adverse consequence which we have no reason to expect but which might conceivably occur. Equally, we should not withhold palpable benefits to future generations because they cannot consent to them of because they might regret or resent our legacy.

To see the truth of this we should start on familiar territory.

Decisions for Those Who Cannot Consent

The suggestion[18] that it might be wrong to do something to or for children because they are not in a position to consent is simply absurd. If decisions could not be made for children unless and until they could consent to those decisions themselves, they would never grow up not to be children. Indeed, they would not live long at all. All sorts of decisions are routinely made for children. Their parents (usually) or guardians or those charged with their care (hopefully) dress them, feed them, talk to them, play with them, hug and kiss and cuddle them, sleep with them, eat with them, travel with them, and educate them. Less unproblematically, they indoctrinate them in religion and other prejudices, expose them to dangers such as carrying them on bicycles, in cars, and on airplanes, cross busy roads with them and sometimes let them cross busy roads or play at the edge of busy roads by themselves, take them for fast food, buy them, or let them eat, or prepare for them, or let them prepare all sorts unsuitable foods rich in cholesterol, sugar, salt, unhealthy fats, etc. Some of these are part of every child's upbringing. How did we allow this state of affairs to occur?

In medical contexts where the consent of a patient cannot be obtained or is simply unavailable, perhaps because that patient is a child or otherwise supposedly incompetent, or because they are temporarily unconscious, someone else consents on their behalf. This is often called "presumed consent" or "substituted judgment" or "proxy consent."

I suggest, however, that the reason why it is right to do what presumed consent or substituted judgment or proxy consent seems to suggest in these cases is simply because treating the patient in the proposed ways is in his best interests and to fail to treat him would be deliberately to harm him. It is the principle that we should do as little harm as possible that justifies treating the patient in particular ways.

The justification for treatment is not that the patient consented, nor that she would have, nor that it is safe to presume that she would have, nor that she will consent when she regains consciousness or when, on ceasing to be a child, she becomes competent, but simply that it is the right thing to do, and it is right precisely *because* it is in her or his best interests.

That it is the "best interests" test that is operative is shown by the fact that we do not presume consent to things that are not in the patient's best interests, even where it is clear that he would have consented. We do not usually mutilate patients who have expressed strong desire for mutilating operations, for example. That we do not, except where we believe it to be in the patients' best interests, amputate healthy limbs of patients suffering body dysmorphic disorder[19] shows that consent is not sufficient justification for all interventions.

If we do not give beneficial treatment to patients who have refused it, say by advanced directive, we perhaps do not do so because we believe this would constitute an assault and a violation of their will. But it is not a violation of someone's will, nor is it an assault to give a treatment they have not refused, the withholding of which would constitute an injury.

The reason it is not a violation is not because they have consented in some notional or fictional sense, but because it is the right thing to do. And if we seek the reason why it is the right thing to do the answer is that to fail or omit to do it would injure the patient. It is the infliction of that injury, by act or omission,[20] that would constitute the violation or assault.

So where, in medical contexts, we act in the best interests of patients who cannot consent, we do so, I suggest, because we rightly believe we should not harm those in our care or even those whom we could affect by our decisions, and not because some irrelevant person or the law has constructed a consent.

So much for medical contexts. If we consider again the case of profiling babies at birth or for that matter all the other decisions that parents and others make for or on behalf of children, we can see that the best interests test is not really helpful either. This is because, whereas it is in the best interests of children that they are fed, clothed, educated, and many other things by competent individuals whether adult or not, this does not justify any particular decision. Indeed, of all the ways a particular parent may, let us say, choose to feed, clothe, talk to, or travel with a particular child, it is unlikely that any would pass the *best* interests test and many would scarcely count as in the child's interests at all.

So, our question then must be: what justifies the imposition of a "best interests" criterion—as, for example, in the case of a decision to consent on a child's behalf to an appendectomy or to dental care, or to take a child in a baby seat on a bicycle or to McDonald's for lunch or to the clinic for an MMR vaccine? The latter is of course added to the list because whereas MMR vaccination is in the public interest because of herd immunity and the "free-rider" effect, it is unlikely to be in the interests of any particular child to be subjected to even the remote risk involved in MMR[21] in a context where most children are being vaccinated. The answer seems to be "nothing." In other words, there really is no feature of decision making on behalf of children which justifies or rather requires the imposition of a "best interests" test rather than of simply a requirement that the decision taken on behalf of children must not be grossly against their interest or manifestly dangerous. Here the standards of danger must be pretty high or fatty foods and bicycle rides in city traffic would be permanently off the menu.

The bottom line seems to be that if we are to permit parents or guardians care and control, in short, if family life is to be protected, huge latitude must be allowed to parents in decision making on behalf of children. There seems to me no reason why that latitude should be denied in the case of genetic testing or profiling at birth or in childhood. Add to this the powerful reasons on both sides of the argument to add to the lack of clarity and the safe side must surely remain with parental discretion.

An important argument made in relation to a child's "right" not to know[22] concerns the removal of the possibility of the child making his/her own decision whether or not to undergo medical treatment at a future time. In the context of genetic testing, it has been said, for example, that

> [t]esting during childhood removes the individual's future autonomy and confidentiality, and may cause damage to their self-esteem and future interpersonal relationships.[23]

Any supposed "right" of children to make up their own minds about having such a test when they are older must, however, be balanced against claims that the failure to test deprives children of the "right" to grow up in an atmosphere of openness and understanding of their situation and a "right" or interest not to form unrealistic hopes and plans about the future. These might include the right to make informed plans and decisions about, for example, rational education prospects (how long a period of preparation for a career would be rational?),

rational career prospects, rational marriage plans, and rational timing of children.

This is not to say that it is obviously in a child's interests to be tested. Indeed, there may be situations when testing is manifestly contrary to a child's best interests. The point to be made, however, is that such interests cannot be determined solely by reference to a child's autonomy interests.[24] Such cases raise real questions about where the balance of a child's best interests lie and consideration must be given to factors other than autonomy that may determine whether a child's medical/genetic status is to be known.

One clear conclusion here is that there is no sense in which a decision not to test children, even for late-onset conditions, protects their autonomy, whereas a decision to test violates it. Again autonomy walks both sides of this street and any such decisions must be based on a calculation of the best interests of the child and on whether or not there are sufficiently powerful and clear grounds to override the parent's presumptive claim to be the guardian of such interests.

If decisions must be taken on behalf of children, the presumption must surely be that health screening is primarily a matter for parents. This and most societies operate rightly with the view that unless and until it can be shown that the presumption that parents are the best guardians of their children's interests must be overturned in order to protect children from serious harm, decisions about most things concerning children should remain with parents.[25]

It is true that to raise a child in the knowledge that he or she will develop a dreaded familial disease may cause additional harm, but there is no reason to assume that this must be so. Unless there is strong reason to suppose that serious harm will result, which I doubt, the principle of nonmaleficence cannot be invoked. The psychological impacts, for example, of being informed of a familial disease are contested.[26] Furthermore, it should be noted that even where the principle of nonmaleficence is appropriately invoked, that is, where disclosing information about the child's health status is likely to do more harm than good, it is always a further and separate ethical question as to whether disclosing information the results of which may not be in the child's overall interests is wrongful. Many people believe that bringing children up in a religion, for example, is strongly against the child's interests, but it does not follow that this practice must be prevented. It would be difficult to defend a diet of so-called "junk food" or of pulp fiction or of the tabloid press, or of game shows, "soaps," and reality television. We tend to forget how much of what happens to children as a result of parental choice or at least of parental "consent" or acquiescence is scarcely describable as either in the child's best interests or indeed even in the child's interests at all. We should be cautious about moving from plausible judgments about the interests of the incompetent to drastic conclusions as to what may or may not be permitted. This caution has been singularly lacking in the literature invoking child welfare and we should all be reluctant to consent to being controlled by those who believe the enforcement of morality is a first duty rather than, perhaps, a last resort.

6 | Disability and Super-Ability

It is a paradox that many of those who oppose human enhancement or indeed who oppose reproductive choices that might influence their children or future generations for the better, that might, in short, enhance, do so partly out of fear. This fear takes a number of forms. One is fear that because some may wish to use enhancement or reproductive technologies to avoid disability or to make better children this will lead to discrimination against people with disabilities. Another is that exercise of these sorts of choices may somehow lead to denying people the freedom to have unenhanced children or even children with disabilities if that is what they want to do. In this chapter we will examine arguments that suggest that eliminating disability is an illicit form of discrimination against people with disabilities. If it is, then it is possible that enhancement might be an even more dramatic form of the same wrongful discrimination.

The idea that selecting against disability somehow constitutes a form of discrimination against those with disabilities is the most plausible[1] of the arguments adduced against reproductive liberty and the exercise of the democratic presumption in considering the question of whether it is legitimate to attempt to make better people.

A thesis of this book is that all persons are equal and none are less equal than others. No enhancement however dramatic, no disability however slight, or however severe, implies lesser (or greater) moral, political, or ethical status, worth, or value. This is a version of the principle of equality.

However, the following three fallacies occur in many discussions of the ethics of choices concerning disability which suggest that reproductive liberty or enhancements involve the denial of the principle of equality; they all involve misunderstandings concerning the equality principle.

- Choosing to repair damage or dysfunction or to enhance function implies either that the previous state is intolerable or that the person in that state is of lesser value, or indicates that the individual in that state has a life that is not worthwhile or not thoroughly worth living. None of these implications hold.
- Exercising choice in reproduction with the aim of producing children who will be either less damaged or diseased, or more healthy, or who will have enhanced capacities, necessarily violates the principle or equality. It does not.
- Disability or impairment must be defined relative either to normalcy, "normal species functioning," or "species-typical functioning." It is not necessarily so defined.

The rest of this chapter is devoted to exposing these fallacies.

Is it morally wrong to attempt to eradicate or indeed to minimize disability? Does such an attempt constitute discrimination against people with disabilities?[2] These questions are of very general importance. They are also intrinsic to the debate about enhancement since, as I have argued, there is a continuum between treating dysfunction and enhancing function. These same questions are of course also questions about enhancement, not simply because of the continuum between therapy and enhancement but because enhancing the functioning or welfare of persons with disabilities will pro rata minimize, or even possibly eradicate, the disability. Moreover, as will be obvious, the very same interventions which in injured, sick, or impaired individuals can help to restore function may, if applied to healthy people or individuals within a normal range of functioning, actually enhance normal or normally healthy functioning.

The question of whether attempts to eradicate or minimize disability or indeed to treat illness or injury constitute some sort of discrimination against people with injuries, illness, or disabilities, or whether such attempts imply that the sick or impaired, injured or disabled are somehow unacceptable as people or citizens is of course of crucial importance.

Discrimination

To address the question of possible discrimination against people with disabilities we will first need to distinguish two prior questions:

- Is it wrong to prefer a nondisabled person to a disabled one?
- Is it wrong to prefer to produce (or even to prefer to be) a nondisabled individual rather than a disabled one?

These questions are often confused. Without arguing for it I will here simply assert and accept that the answer to the first question is an emphatic "yes." I will assume that it is wrong to show preference for a nondisabled person over a disabled one in any way that denies that both are equally entitled to the same concern, respect, and protection as is accorded to any citizen. For such preference would imply that the disabled person was somehow less than an equal in moral and/or political terms.[3] We are of course only talking of preferences which carry this implication. Thus, a preference to be operated on by a brain surgeon who is not suffering from Parkinson's disease is not a preference which assaults the dignity of those with this damaging condition.

The crucial question is the second one. Is it wrong to prefer *to produce* or to prefer *to be* a nondisabled individual rather than a disabled one?[4] Would such a preference violate the principle of equality implicit in the first question and accepted by this author?

Should We Attempt to Prevent Disability?

To try to resolve the question of the ethics of attempting to prevent disability, we might well begin by examining a series of questions which look at the different means by which we might make this attempt to see what light this throws on the ethics of different courses of action. I will list them first and then discuss each briefly.

Is it wrong to prefer to produce a nondisabled child and attempt to achieve that preference

- by wishing and hoping?
- by behavior modification?
- by postponement of conception?
- by interventions, therapeutic or enhancing (including gene therapy)?
- by selecting between preimplantation embryos?
- by abortion?

Is it wrong to hope and wish that one's child[5] will not be disabled in any way? I mean is it wrong to wish this full-heartedly, knowing that the making of the wish, perhaps publicly, might be thought by some to constitute some sort of "criticism" or devaluing of people with disability? And if it is not wrong to wish for the healthy child in this full-hearted sense, would it be wrong, if we had the power, to play fairy godmother to ourselves and grant our wish? Well, of course, in wishing that my child will not be disabled I am not committing myself to the view that should it nonetheless be born disabled it would then have a life not worthy to be lived. I am saying that it is better that my child be not disabled, but not that if it is disabled, a nondisabled child is a better child. These are, of course, different senses of "better."

Now reverse this suggestion. What would we think of someone who wished that her child be born disabled rather than healthy? Would a decent person wish to have a disabled child? Would a decent person try to grant their own wish? Some apparently decent deaf people do in fact wish their children to be deaf like them and resist therapies to improve the hearing of their children.[6] They suggest that there is a distinctive deaf culture which is in some senses better than that available to those with hearing. Perhaps the test here is whether or not we would feel a deaf parent justified in deliberately taking steps to ensure that any future child would be conceived with deafness genetically guaranteed in order to ensure that it had secured to it the benefits of deaf culture. Would we accept that this was a morally neutral piece of "medical" intervention (perhaps like male circumcision—if that is in fact morally neutral?) or rather a deliberate disabling or "mutilating" act? And suppose such a parent were fortunate enough to succeed in having the deaf child they desired, but suppose also that this form of deafness could be successfully removed with medical treatment. Would the parent be justified in denying the child the relevant therapy? We will return to this question in due course.

Suppose a woman is contemplating attempting to conceive and she knows that if she continues her consumption of alcohol or cigarettes, she is likely to disable her child in some way. Should she stop these things and modify her behavior so as to maximize the chances of her child not being thereby injured? Again I believe she should and that she would be wrong not to: wrong because to fail to modify her behavior would be to deliberately risk injuring her child.

Let's look at postponement of conception and gene therapy together. The same arguments would apply. Suppose a woman could, by therapy in utero, remove disability; should she do so? It is difficult to imagine how one might defend the actions of a mother who would not treat illness or disability, or impairment in her child if the required process carried an acceptable level of risk given the benefits. Now suppose she is affected by a condition which will disable any child she conceives now, but if she postpones conception and takes a course of treatment she can conceive normally. Ought she to postpone conception in order to avoid having a disabled child? Again it looks strongly as though she should act so as to avoid having a disabled child. This will be true even where the child she will have after treatment will not be "made better" by the treatment and will not be the same as the child she would have had but for the treatment. In other words, we still believe that she would be right to act so that she has a healthy rather than a disabled child although it involves choosing between possible children rather than making one child well.[7]

Suppose, as is in fact normal practice, a woman using in vitro fertilization (IVF) techniques has had five eggs fertilized and now wishes to use her embryos to become pregnant. Normal practice has up to now been to insert two embryos in the hope of achieving one healthy birth.[8] Suppose preimplantation screening had revealed two of the embryos to possess disabilities of one sort or another. Would it be right to insert the two embryos with disability or to choose randomly which embryos to implant? What would justify the suggestion that it would not be wrong to implant the two embryos with disability and thus deliberately choose to have a disabled rather than a healthy child?

The Justification Is Not the Reason

But notice that in the justification for declining to implant particular embryos disability plays no role at all. The mother would be entitled to decline to implant even healthy embryos. Her *reason* for not implanting these particular embryos may be that they are disabled, but her *justification* is in terms of her entitlement to decline to implant any at all, disabled or not. For example, a woman's reason for choosing an abortion may be that she does not want to run risks to her health but the justification must be that the moral importance, the moral status, of the embryo is not such as to require those risks of her.[9]

The conclusions we have so far reached are unsurprising. Of course, they are predicated on the idea that disability is somehow disabling and therefore undesirable. If it were not, there would be no motive to try to cure or obviate disability in health care more generally. If we believe that medical science should try to cure disability where possible and that parents would be wrong to withhold from their disabled children cures as they become available, then we will be likely to agree in our answers to the five questions raised on p. 89, above.

What Is Disability?

The answer to the question, What is disability? is of more than semantic importance. I have defined disability as a condition that someone has a strong rational preference not to be in and one that is moreover in some sense a harmed condition.[10] I have in mind the sort of condition for which if a patient presented with it unconscious in the emergency room of a hospital and the condition could be easily and immediately reversed, but not reversed unless the doctor acts without delay, a doctor would be negligent were she not to attempt reversal. I call this "the emergency-room test." Or, one in which, if a pregnant mother knew that it affected her fetus and knew also she could remove the condition by simple dietary adjustment, then to fail to do so would be to knowingly harm her child.[11] It will be clear that on this account of disability it would not be a disability to be born a Jew or a gipsy or black in Germany in the 1930s because whereas someone might well have a strong rational preference not to be in any of these "conditions" in the stipulated circumstances, they are none of them in any sense "harmed conditions." However, as we have seen (in chapter 4), to be born with the possibility of living only a normal life span when others can live substantially longer, and when the individual in question could have been enabled to live longer, would count as being disabled to some degree because rational people would want the chance of a longer life (remember they don't have to accept that chance). Equally, if they were denied such a chance, they would be left in a harmed condition relative to those with longer life expectancy. Short life expectancy would surely meet "the emergency-room test."

On this view a harmed condition is defined relative both to one's rational preferences and to conditions which might be described as harmful. Disability then is defined not relative to normal species functioning but relative to possible alternatives. This is very important because so many of those who write about disabilities not only persist in the fallacious view that disability, impairment, or indeed illness must be defined relative to normal species functioning, or species-typical functioning, but seem unable to contemplate clear alternative accounts. Normal species functioning cannot form part of the definition of disability because people might be normal and still disabled.[12] Suppose that, due to further depletions to the ozone layer, all white-skinned people were very vulnerable to skin cancers on even slight exposure to the sun, but brown- and black-skinned people were immune. We might then regard whites as suffering substantial disabilities relative to their darker-skinned fellows. And if skin pigmentation could easily be altered, failure to make the alterations would be disabling. We will return to the issue of enhancements later. For the moment it is sufficient to note that in such circumstances whites might have disabilities relative to blacks even though their functioning was quite species-typical or normal.

It is of course difficult to spell out exactly what one would and should call a "harmed condition." Harms can be quite slight but still be harms. I have suggested that a harmed condition is one for which if a patient was brought unconscious into the ER department of a hospital in such a condition and it could be reversed or removed, the medical staff would be negligent if they failed to reverse or remove it. So, although the loss of the first joint of the little finger would be a small harm to bear, if someone came into hospital with the little finger severed at the first joint and it could be sewn on again, the staff would be negligent not to do so; they would have harmed the patient by failing to restore the finger.

These points are crucial because it is sometimes said that while we have an obligation to cure disease—to restore normal functioning—we do not have an obligation to enhance or improve upon a normal healthy life, that enhancing function is permissive but could not be regarded as obligatory. But what constitutes a normal healthy life is determined in part by technological and medical and other advances (hygiene, sanitation, etc.). It is normal now, for example, to be protected against tetanus; the continued provision of such protection is not merely permissive. If the AIDS pandemic continues unabated and the only prospect, or the best prospect, for stemming its advance is the use of gene therapy to insert genes coding for antibodies to AIDS, I cannot think that it would be coherent to regard making available such therapy as permissive rather than mandatory.

This is not an exhaustive definition of disability but it is a way of thinking about it which avoids certain obvious pitfalls. First, it does not define disability in terms of any conception of normalcy. Second, it does not depend on post hoc ratification by the subject of the condition: it is not a prediction about how the subject of the condition will feel. This is important because we need an account of disability we can use for the potentially self-conscious—gametes, embryos, fetuses, and neonates—and for the temporarily unconscious, which does not wait upon subsequent ratification by the person concerned.

One of the best and most sensitive discussions of disability occurs in Jonathan Glover's wonderful new book *Choosing Children*.[13] We will be returning to Glover's ideas later but for the moment it is important to note a reservation he has over my own account of disability just discussed:

> For Harris, anyone in any disadvantageous condition is harmed by whoever caused him or her to be in that condition. Because all of us have some disadvantages, we have

> all been harmed by the parents who caused our birth. To Harris, Philip Larkin's gloomy conception of the role of "your Mum and Dad" must seem a huge understatement, as Larkin thought they started to "fuck you up" only after birth. I salute John Harris's cheerful espousal of the even gloomier view, but will not follow him.[14]

I enjoy salutations, especially from Jonathan Glover, but he has not got me (or it: the argument) quite right in this case. True, on my account our parents are causally responsible for the genetic condition we are born in; how could it be otherwise? In that trivial sense they have harmed us, if harms attach to our genetic constitution, as some almost always do. However, they are surely not *morally* responsible: they haven't deliberately "fucked us up" to the extent of the harms, unless they were, first, aware that they were likely to transmit those harms and, second, aware of a better alternative child, or a better possible alternative child, and could, realistically, have produced that child instead.

Here there is not space to argue the point, but there are powerful reasons to respect the "right to found a family" or, for those with an interest-based theory of entitlement, to respect the impulse and interest in founding a family. So long as the life of the child you will produce will be highly probably worth living, it is in that child's interests to be born and hence you will benefit and not wrong that child[15] by bringing it into existence. So long as it is not possible to produce a healthier, and probably happier, alternative child there are still good moral reasons to produce children so long as their lives are predictably well worth living. I have argued that, although there are good reasons to avoid bringing children into existence who will have lives that are predictably harmed in any way, the interest in having a child is a powerful one.[16] So long as it is in that child's interests to be born, which means so long as that child will predictably have a life worth living, then that child (though it may be born in a harmed state) is not wronged by being brought to birth. Unless of course that child is wronged because someone could reasonably have made her life more worthwhile and deliberately failed to do so. Indeed, if, despite low probability of this happening, a child turns out to have a life that is not worth living, then still she is not wronged because the *expected* utility of her life is positive. In *Wonderwoman and Superman* I suggested that

> [w]e have recognized the powerful desire and the strong interest that people generally have in having children. Just as this desire should be exercised responsibly, we should also be careful not to frustrate it without good reasons.
>
> If children are wanted, it is better to have healthy children than to have disabled children where these are alternatives, and it is better to have children with disabilities than to have no children at all.[17]

The proviso of course is that the children with disabilities will have lives that are worth living. Where this is reasonably predicted to be the case, even if disabilities or other harms are also predicted, it is in the child's interests to be brought to birth.

We must now turn to the last two questions on our list, which concern selection between embryos where the embryos not selected are not implanted and abortion for disability.

Disability and Discrimination

It is often said by those with disability or by their supporters (who should of course include us all) that abortion for disability, or failure to keep disabled infants alive as long as possible, constitutes discrimination against the disabled as a group, that it is tantamount to devaluing them as persons,[18] to devaluing them in some existential sense. Alison Davis identifies such talk with utilitarianism and comments further that "[i]t would also justify using me as a donor bank for someone more physically perfect (I am confined to a wheelchair due to spina bifida) and, depending on our view of relative worth, it would justify using any of us as a donor if someone of the status of Einstein or Beethoven, or even Bob Geldof, needed one of our organs to survive."[19] This is a possible version of utilitarianism of course, but not I believe one espoused by anyone today. On the view assumed here and which I have defended in detail elsewhere,[20] all persons share the same moral status whether disabled or not. To decide not to keep a disabled neonate alive no more constitutes an attack on the disabled than does curing disability. To set badly broken legs[21] does not constitute an attack on those confined to wheelchairs. If choosing to create children without disability is some sort of slight against people with disabilities, then why isn't treating curable illness some sort of slight against (or unfairness to) those with incurable disease? To prefer to remove disability where we can is not to prefer nondisabled individuals as persons; to treat illness where we can is not to exhibit a preference for the healthy as persons—it is not the demonstration of what we might call an "existential" preference. To reiterate, if a pregnant mother can take steps to cure a disability affecting her fetus, she should certainly do so, for to fail to do so is to choose to handicap her child. She is not saying that she prefers those without disability as persons when she says she would prefer not to have a disabled child.

I have said that the decision to abort for disability no more constitutes an attack on the disabled than does curing disability. I believe this to be true, but its truth depends upon the moral status of the fetus being the same as that of the embryo and the gametes and not the same as human individuals who have developed the capacity to value their own existence—persons properly so-called. This argument I have developed in detail on a number of occasions[22] and for present purposes I will simply summarize it, for the conclusions I want to draw depend not upon the truth or cogency of the argument but rather on the shared morality of which the argument (whichever side of it is taken) is a part.

The Moral Status of Human Individuals

If it is wrong deliberately to implant, say, severely disabled preimplantation embryos rather than healthy ones, or if it is permissible to abort a fetus in order to save the life of or preserve the health of its mother, then there must be a morally significant difference between the moral status of the embryo and the fetus and that of the mature adult. Most people believe that there is some such difference and no one could accept abortion under any circumstances (save randomly, where either the mother or the fetus can live but not both[23]) without accepting some version of the thesis that there is a morally significant difference between the fetus and its mother. My own suggestion is that the moral status of the embryo, the fetus, and indeed any individual is determined by its possession of those features which make you or me morally more important than cats or canaries. At no stage of its development does the human embryo or fetus possess features that relevantly distinguish it from cats and canaries, save two. They are species membership and potential. Species membership, however, is devoid of moral significance; species preference is, like race or gender preference, simply a prejudice. Potential is no more helpful: whatever potential is possessed by the human embryo is also possessed by the unfertilized egg and the sperm and so the argument that we have an obligation to realize human potential becomes the exhausting and unattractive ethic of maximal procreation.[24]

It is sometimes claimed that because fertilization occasions the formation of a new personal identity the gametes lack the potential that the new individual, the zygote, possesses. But personal identity has nothing to do with this argument. Consider that the zygote has a potential, namely the potential to become a glorious full person like you and me. Well, *something* or *-things* must have had the potential to become the zygote. If the zygote has value because of its potential to become a person, then whatever has the potential to become a zygote shares whatever importance the zygote has in virtue of its potential to become a person.

On my account, persons properly so-called are individuals capable of valuing their own existence. This view, making features which depend upon the capacity for self-consciousness and a minimal rationality central, is like that of John Locke and many philosophers since.[25] The important point is that a view like this distinguishes persons with such capacities from embryos, fetuses, and neonates not on the grounds of presence or absence of disability but on the grounds of presence or absence of the capacities that make for moral significance.

So much for the summary of the argument. It is of course just that: a summary and not itself an argument. It will not convince those who disagree. The important point for present purposes, and one which bears repeating, is that the moral difference between the embryo, the fetus, and the neonate depends at no point and in no part on possession or absence of disability. If preimplantation embryos may be destroyed, if fetuses may be aborted or neonates allowed to die, the justification must be in terms of their moral status. Those, like myself, who argue that the neonate is not morally significantly different from the unfertilized egg on the one hand, and from cats and canaries on the other, owe an account of why this is so.[26] Those who deny it also owe an account of why it is not so. I do not believe that disability is relevant to any of these accounts, whether from one side of the argument or the other.

That this is so can be seen from the other side of the argument. Alison Davis is a prominent defender of the rights of the disabled as we have already noticed. But her defense is from a "pro-life" perspective. On the pro-life view defended by Davis, abortion is never justifiable, not even to save the life of the mother, or however disabled the fetus is and the child which it will become will be. On this view the mother's life is of no more importance than that of the fetus and both must be equally respected. On such a view abortion will be ruled out whether or not the fetus is disabled. In particular, and dramatically, abortion is ruled out if the fetus will have no chance of being carried to term and even if the mother will die if it is not aborted. The individual who takes a view like this must argue that there is no justification for killing one morally important being to save the life of another where both are of equal status.[27]

In the justification of a position like that of Alison Davis, handicap and disability play no role at all. Nor do they, nor need they, in the justification of views from the other side. The moral status of no individual, whether embryo, fetus, neonate, human adult, cat, or canary, is altered one jot by the existence or absence of disability.

A Life Not Worth Living

The notion of a life not worth living or a life not worthy to be lived is often invoked in discussions of disability. I do not believe that this notion plays any role at all in the justification of decisions not to produce a disabled child in the sorts of circumstances identified above. No one, I believe, would say that the lives of most people with disability are not worth living. All that is claimed is that it would be better not to have a disability: if it would not, then it would not be a disability. Certainly, none of the decisions not to produce a disabled individual depend on making the judgment that that individual's life would be not worth living or that disabled individuals are not worthy to live.[28]

If, say, I were to lose the use of my legs and become disabled to that extent (or for that matter to succumb to some mental disability or illness), I would regard it as a severe misfortune, perhaps even a tragedy, but I would not have become less valuable in the "existential" sense or less morally important on that account. My life would be no more or less subordinate to those not disabled than it was before. I hope I would find life still worth living and, if I did, that this would not change my views about the rightness of wishing for a child who was not paraplegic (or, for example, brain damaged[29]) nor on the rightness of acting on such a wish if I were able so to do.

I have a rational preference to remain nondisabled, and I have that preference for any children I may have and indeed for humankind[30] in general. To have a rational preference not to be disabled is not the same as having a rational preference for the nondisabled as persons.

I have tried both to explain that we should try to eradicate disability and to show why this view in no way implies or involves discrimination against handicapped people. I believe that the rightness of choosing not to produce disabled individuals in the list on p. 89 is so clear that it would be an embarrassment to any view to have to differ.[31] Intuitions over the cases of abortion or infanticide are likely to be more divergent. I hope I have said enough to show that disagreements over the legitimacy of aborting disabled fetuses (permitted by legislation in most countries of the European Union) are disagreements about the legitimacy of abortion and not about attitudes to disability. Infanticide, more problematic certainly, involves an extension of the same argument. Those who argue that it should be permitted do so because they see no morally relevant difference between neonates and fetuses; those who disagree must show what morally relevant differences obtain. But again these differences will have nothing to do with the presence or absence of disability.

Reasons and Justifications

A practical example can help us here. The United Kingdom *Human Fertilisation and Embryology Act* 1990 permits abortion to prevent fetal handicap up to birth. Since there can be no difference, moral or physical, between a fetus killed "en ventre sa mère" at 27 weeks, 6 days, 23 hours, and 59 minutes, and one killed as a neonate five minutes later, it cannot be a great leap of morality to at least question whether, if the abortion is morally legitimate, the infanticide might not also be morally legitimate. The same minute difference also shows that the operative justification for killing the fetus but not the neonate had in fact nothing to do with the level of disability or even the presence of disability, but rather involved *prior* acceptance of the legitimacy of abortion rather than infanticide (however ethically confused such acceptance might be). That is to say, the moral justification for accepting fetal handicap as a reason for abortion right up to term depends upon the judgment that abortion is permissible, justifiable, in a way that infanticide is not; justifiable, that is, in a way that admits of the consideration of a reason. The judgment that abortion can be justifiable is thus prior to the consideration of particular reasons for abortion in a given case. The fact of handicap or disability plays no role in that prior judgment. The reasons for abortion in the particular case are only considered because a judgment has *already* been made that abortion is permissible. In that prior judgment disability plays no part. This is shown by the fact that in the case of infanticide for disability, though the reason (the disability) is as strong as before, the justification for infanticide is clearly felt to be absent.

I have attempted to explain how abortion and even infanticide for disability are extensions of the legitimate, perhaps imperative, ethic of combating disability, and, in particular, that it is on a continuum, unbroken by any distinction of ethical significance, with attempts to eradicate disability by wish fulfillment or by the implantation of nondisabled embryos. The significant distinction is that persons are morally important in a way that no other sorts of creature are. Persons can be unjustly discriminated against in ways that nonpersons cannot. The only reason not to select healthy embryos or not to abort disabled fetuses is not because this would be unjust to the embryos or fetuses but because it might be offensive to those disabled persons who felt it either disvalued them as persons or threatened their equal standing in some way. The fear is here that the selection of embryos that do not have disabilities is somehow an expression of what Jonathan Glover has called ugly attitudes toward people with disabilities.

Ugly Attitudes toward People with Disabilities[32]

There are two types of indirect costs to the disabled of combating disability in the ways outlined here. There are what might be termed "subjective costs"; that is, costs in terms of the feelings of being disvalued aroused by attempts to prevent the existence of people *like them*. Then there are also more "objective" indirect costs in terms of the ways in which a policy of combating disability in the ways outlined might play some causal role in leading people with disability to be less valued by the community at large and consequently become the victims of discrimination.

While both these sorts of costs should be limited so far as is possible, they each present different problems. So far as what I have called subjective costs are concerned, if, as I have suggested, they are irrational or misconceived, then we must ask a pertinent question. Can there be an obligation to protect some people against inappropriate feelings that they may have at the cost of real harm and possibly pain and suffering to those who must continue to be born with disabilities just so that existing disabled people be protected from inappropriate feelings of rejection? As we have noted, these feelings might be and perhaps are aroused not simply by attempts to eradicate disability in future people but by therapies for disability in their fellows.

The second, "objective," sorts of harm are easier to deal with. We should certainly combat such unjust and unjustified discrimination against disabled people by all means at our disposal, including legislation.

Jonathan Glover has provided an extensive discussion of what he calls "some ugly attitudes towards people with disabilities and disorders."[33] Glover notes that the "decision not to have a child with a disability may be an expression of ugly attitudes towards people with disability. But equally it may not." Minimizing the danger of people with disability suffering any damage from ugly attitudes is obviously the objective of any decent person. While emphasizing that he accepts that, other things being equal, it is good if the incidence of disabilities is reduced by parental choices to opt for potentially more flourishing children, Glover outlines three strategies for achieving this:

> To do this, we need to send a clear signal that we do not have the ugly attitudes to disability. It is important to show that what we care about is our children's flourishing: that this, and not shrinking from certain kinds of people, or some horrible project of cleansing the world of them, is what motivates us. To think that a particular disability

> makes someone's life less good is not one of the ugly attitudes. It does not mean that the person who has it is of any less value, or is less deserving of respect, than anyone else.
>
> There are two ways we can show this. One is by making the comparison with other medical programmes. We want to defeat cancer, not because we lack respect for people with cancer and want to rid the world of them, but because of what cancer does to people....
>
> The second way of reducing the harm requires us to see one of the implications of our view. We want parents to have the choice of having a child without a disability because disabilities reduce the chance of flourishing. But disability is only one way in which flourishing is impaired. Poverty, bad housing, or child abuse can do so at least as much. If we single out disability among the obstacles to flourishing, the ugly attitudes may seem to be lurking there. We have to take the other obstacles just as seriously.[34]

I have said, and I repeat, that having a rational preference not to become disabled oneself and not to have children with disability is not the same as having a rational preference for the nondisabled as persons.

Many people talk as if the disabled are simply differently abled and not harmed in any way by so-called disability. Deafness is often taken as a test case here. Insofar as it is plausible to believe that deafness is simply a different way of experiencing the world, but by no means a harm or disadvantage, then of course the deaf are not suffering from any disability. But is it plausible to believe any such thing? For example, both Tom Koch and S. D. Edwards sometimes talk as though deafness were not a harm or a deficit.[35]

Imagine a child whose deafness could have been successfully treated saying the following to the parents who denied her the treatment: "I could have enjoyed Mozart and Beethoven and dance music, and the sound of the wind in the trees and the waves on the shore; I could have heard the beauty of the spoken word and in my turn spoken fluently but for your deliberate denial." In response Koch suggests that "one may acknowledge the joy [that these things] bring others without insisting that the inability to perceive them is a harm or a deficit. After all, many persons are 'deaf' to the pleasures of classical music (or jazz, or reggae, or rap, etc.) and yet none assume their limits of comprehension reflect a deficit or harm." In similar vein Edwards suggests: "I suppose it may be said that a moderately intellectually disabled person misses out on those dimensions of experience which require considerable intellectual acumen.... And it may be said of those without musical ability that they miss out on *that* dimension of human experience."[36]

But to be "deaf" to the pleasures of classical musical is to be deaf in inverted commas, not really deaf. Musical taste can be educated, but not so hearing for the profoundly deaf. Edwards's point is rather different. The intellectually disabled do miss out on some dimensions of experience which are closed to them in the way that music is closed to the deaf. And this is a disability. It may be a moderate or even a slight disability. Life may well be not only tolerable but truly excellent with such a disability.[37] Like the loss of the end joint of a little finger, the point is not that life is not worth living without such things but that we have reasons not to start out in life with any unnecessary disadvantages, however slight.

Would the statement "I have just accidentally deafened your child; it was quite painless and no harm was done so you needn't be concerned or upset!" be plausible? Or suppose a hospital were to say to a pregnant mother, "unless we give you a drug your fetus will become deaf, since the drug costs £5 and as there is no harm in being deaf we see no reason to fund this treatment." But there is harm in being deaf and we can state (and have stated) what it is.

We are now in a position to return to the main question: is it better to avoid bringing people with disabilities into existence where possible?

Neither levels of impairment, nor suffering, nor normal functioning are the issue when it comes to reproductive choice. We are asking the wrong questions about disability if we think that the ethics of reproductive choice turn on degrees of disability or the subjective experience of disability. If we ask "what would justify the prevention of this life or a life like this?" or "what impairments or levels of impairment or deviation from the norm would justify abortion or selection of embryos?" we beg the question. This is because if we ask such questions, the loss of a little finger or even deafness seem doubtful candidates for reasons to prevent the existence of a person who will nonetheless have a good life.

Concentration on justifying abortion fogs the issue with irrelevant prejudices about what it takes to justify abortion. Discussions turn often on what sorts of features of existence or disabilities would justify abortion. And this is seen (wrongly) as the same question as asking what features of existence are so bad that it would be better never to have lived rather than live in such a condition. And of course we never know in advance what our embryo would grow up to be like; all we have are statistics about embryos like him or her. An example is the following:

> I feel that it (testing) gives the ordinary person (my neighbour, my friends and family, the nurses and other medical professionals who may care for me) the impression that I, as a genetically disabled woman, have a duty to abort an "abnormal" foetus. I challenge this. If others see it as my duty to abort a child of mine simply because it is like me, surely according to their logic, I have no right to live either.[38]

Instead we should concentrate on reproductive choices and in particular the least morally problematic of such choices. Let's start with preimplantation genetic diagnosis (PIGD). Suppose a woman has six preimplantation embryos in vitro awaiting implantation. PIGD has revealed that three have various genetic disorders and three seem healthy. Which should she implant? Does she have any moral reasons to avoid implanting those with genetic disorders? Notice two features of this case. The woman is under no moral nor any legal obligation to implant any of the embryos. The decision to implant some or none is entirely within her unfettered discretion. She doesn't have to offer legal, moral, or any other justifications to anyone if she decides to implant none of the embryos. Under English law she may only implant up to three without a special medical reason for implanting more. Which three should she implant? Can she say, "It is a matter of moral indifference whether or not my resulting child has a genetic disorder and therefore I have no reason to select the healthy embryos"? This seems implausible. Since none of the embryos has a right or an entitlement to be chosen rather than the others, since none is a person, nor yet a moral agent, none has begun the sort of biographical life that would give it interests, her choice is relatively free. She has a reason to do what she can to ensure that the individual she chooses is as good an individual as she can make it. She has a reason therefore to choose the embryo that is not already harmed in any particular way and that will have the best possible chance of a long and healthy life and the best possible chance of contributing positively to the world it will inhabit.

If on the other hand she chooses to implant an individual destined to suffer an illness, she will have created that illness and any harm that it will do. This woman has the same reason to select against an embryo with a genetic disease as her sister, who is told that if she conceives immediately she will have a child with a genetic defect but that if she postpones pregnancy and takes a course of treatment she will have a healthy child.

The question we should ask is: what reproductive choices would be legitimate and which, if any, reproductive choices would be wrongful? Before exploring further why this is the appropriate question, we should look at what Tom Koch has to say:

> Finally, the assumption of future harm as a basis for eugenic selection is difficult to justify and difficult to apply to conditions [like] ... ALS/MS (amytrophic lateral sclerosis/multiple sclerosis), familial Alzheimer's, Huntington's chorea, etc.... To eliminate the person who might develop these conditions in midlife or later would be to deprive society at large of people like physicist Stephen Hawking (ALS), former U.S. president Ronald Re[a]gan (Alzheimer's) or singer Woody Guthrie (Huntington's).[39]

While there are obviously different degrees of regret appropriate to the idea of depriving society of some of the individuals identified, the point is that the idea that eugenic selection has any of these effects is absurd. We are here being offered the famous "aborting Beethoven" fallacy.[40] To choose not to have a child with inherited syphilis is not to decide that the world would be better off without Beethoven. It is as senseless to bemoan the fact that we have elected not to create "a Beethoven" as it would be to celebrate the fact that, by practising contraception, we have just prevented the birth of "a Hitler."

Consider the question, should John Harris have been born? (I am sure that a number of people have asked themselves this question.) Suppose my parents had been told in the year in which I was conceived that by postponing conception and taking a liter of orange juice every day for three months they would get a cleverer, healthier, longer-lived child. Had they chosen this "optimizing strategy," three obvious questions arise:

- Would they have done wrong?
- Would they have wronged me or people like me?
- Would I have had any grounds for complaint?

I cannot see that my parents would have done anything wrong or blameworthy had they made this choice. There is no one they would have wronged or harmed. And even if they had chosen to abort "me" rather than postpone conception with the result that "I" never existed, the same would have been true. Had they done so they would not have been depriving the world of anyone with particular features or skills or who lacked particular features or skills. Society would not have been "deprived of" John Harris nor would it have been "protected from" him—he (I) simply would never have existed.[41] Those people who, like me, defend abortion know that as a result both healthy fetuses and some with genetic diseases will never become persons.[42] That does not mean that we deprive society of people like Einstein or Ghandi or people like Stephen Hawking or Woody Guthrie. Nor does it mean that we discriminate against such people nor against people like them.

It might of course be suggested that the mere existence of disabled people is good for society, for the rest of us, that we all benefit from their being among us. Of course we do! This is not at issue. What is fundamental is that, once we have the choice as to whether to create people with disability or not, we have to decide whether we are entitled to impose disability on some in order for others to reap the benefit of their presence among us. This seems doubtful, not least because if such benefits would justify reproductive choices that result in disability, it is difficult to see how they would not justify the deliberate disabling of some existing people to get the same benefits.

Choosing Who Shall Exist

Choosing between existing people for whatever reason always involves the possibility of unfair discrimination because there will, inevitably, be people who are disadvantaged by the choice. Choosing which sorts of people to bring into existence or choosing which embryos or fetuses to allow to become persons can never have this effect because there is no one who suffers adversely from the choice.

My parents were under no obligation to attempt to conceive in any particular month. If they had conceived in any month other than the one in which I was conceived, I would not have existed. Not only are none of my possible siblings, who have been irrevocably harmed by this choice of my parents, complaining, I can assure you that had my parents chosen not to attempt to conceive that month (or had their attempt, if that is what it was, been unsuccessful) you would never have heard me complain about it.

Suppose IVF and PIGD had been available to my parents and I had existed in a petri dish.[43] Suppose my parents had been informed that there were other embryos with a better genetic constitution than mine who would probably live longer and healthier lives and they had acted to optimize the life chances of any children they would have. Would I have had any grounds for complaint? Would that have constituted discrimination against people with my genetic condition? I don't believe so. It is simply a fallacy to think that choosing between preimplantation embryos or choosing to terminate pregnancies of embryos because other embryos would have a better chance in life constitutes unfair discrimination.

Enhancements

This fallacy can clearly be seen if we consider again not the issue of disabilities or impairments but rather the issue of enhancements. Which brings us nicely back to the theme of this book, for enhancements show us precisely why selecting against some embryos or in favor of others carries with it no implication or judgment that the individuals selected against have lives that are in any sense not worth living or not worthy to be lived. Suppose some embryos had a genetic condition which conferred complete immunity to many major diseases—HIV / AIDS, cancer, and heart disease, for example—coupled with increased longevity. We would, it seems to me, have moral reasons to prefer to implant such embryos given the opportunity of choice. But such a decision would not imply that normal embryos had lives that were not worth living or were of poor or problematic quality. If I would prefer to confer these advantages on any future children that I may have, I am not implying that people like me, constituted as they are, have lives that are not worth living or that are of poor quality.

Most disabilities fall far short of the high standard of awfulness required to judge a life to be not worth living. This is why I have consistently distinguished having moral reasons for avoiding producing new disabled individuals from enforcement, regulation, or prevention of the birth of such individuals. This is why I have specifically and repeatedly said (and feel I must say again now) that for those who can only have children with disabilities, having such children may well be morally better, for the parents and for the children, than having no children at all.

7 | Perfection and the Blue Guitar[1]

The man bent over his guitar,
A shearsman of sorts. The day was green.

They said, "You have a blue guitar,
You do not play things as they are."

The man replied, "Things as they are
Are changed upon the blue guitar."

And they said then, "But play, you must,
A tune beyond us, yet ourselves,

A tune upon the blue guitar
Of things exactly as they are."

—Wallace Stevens, *The Man with the Blue Guitar*

One crucial issue at stake in discussions of enhancement is whether the "tune" of change is "beyond us yet ourselves" or beyond us in a way that alters our essential nature, our humanity. There is a strong lobby against changeable tunes. Arguments against human enhancement tend to be conservative, they express suspicion of change, emphasize the virtue of things as they are and acceptance of those things. In "The case against perfection: what's wrong with designer children, bionic athletes, and genetic engineering"[2] Michael J. Sandel makes an eloquent case against enhancement and insists on a position which is in effect "A tune upon the blue guitar / Of things exactly as they are."

Sandel does not follow many popular writers in talking simplistically as if the attempt to "design" children was necessarily a quest for "the perfect child." He is aware that this idea is incoherent. There is no such thing as perfection, not least because there is no agreed or even agreeable account of what human perfection might consist in. There are, however, imperfections aplenty and the attempt to minimize these is Sandel's real target.

He suggests that

> [i]n order to grapple with the ethics of enhancement, we need to confront questions largely lost from view—questions about the moral status of nature, and about the proper stance of human beings toward the given world. Since these questions verge on theology, modern philosophers and political theorists tend to shrink from them. But our new powers of biotechnology make them unavoidable.

This is unconvincing; many contemporary writers are concerned with enhancement in one way or another and to have a view on the enhancement of human beings is to have a view about "the proper stance of human beings toward the given world." Indeed, to have a view about the legitimacy of medicine and medical research, mainstream subjects of inquiry for more than 2,500 years,[3] also involves "the proper stance of human beings toward the given world." As we shall see, there are substantial problems in making sense of the idea of "the given world" when so much of human history has been concerned with attempts to improve upon the given, whether by animal husbandry, farming, plant breeding, irrigation, architecture, or in a myriad of other ways. We will ignore the "given world's" attempts to improve upon itself through an evolutionary process of which (arguably) enhancement is a part.

Sandel starts by considering some examples of enhancement and dismissing many poor arguments that have been leveled against the enterprise of enhancement. Having dismissed what he thinks of, often appropriately, as the poor arguments, Sandel sets out to provide better arguments against enhancement:

> It is commonly said that genetic enhancements undermine our humanity by threatening our capacity to act freely, to succeed by our own efforts, and to consider ourselves responsible—worthy of praise or blame—for the things we do and for the way we are. It is one thing to hit seventy home runs as the result of disciplined training and effort, and something else, something less, to hit them with the help of steroids or genetically enhanced muscles. Of course, the roles of effort and enhancement will be a matter of

> degree. But as the role of enhancement increases, our admiration for the achievement fades—or, rather, our admiration for the achievement shifts from the player to his pharmacist. This suggests that our moral response to enhancement is a response to the diminished agency of the person whose achievement is enhanced.

This claim seems doubtful for a number of reasons. The first is that it draws an odd and unexamined distinction between self-improvement which involves greater effort and that which involves lesser effort. In short, it is not rational to think that only effortful rather than effortless superiority counts.

I may try, for example, to improve my health by strenuous exercise and improve my chances of longevity by practising extreme dietary restriction. Both of these require effort, willpower, and continual if not continuous application and determination. These clearly meet the Sandel test. But suppose, as already noted, I prudently take aspirin and statins to minimize cholesterol and protect the cardiovascular system and I follow the attractive, but also beneficial, "Mediterranean diet"? These do not require much effort on my part, but surely I am responsible for these acts of mine and their effects. I hope the admiration I deserve for taking care of myself does not shift from me to my pharmacist or to my purveyors of fresh fruit, vegetables, red wine, and olive oil. I am no less the agent in these choices than in my choice of exercise or dietary restriction. The same goes for practice and steroids. Even steroid-enhanced athletes need the maximum degree of practice and training to have a chance of winning both against their competitors who don't also take enhancing drugs and against those who do. Isn't it rather that effort is still needed but the athletes start from a higher threshold? Authentic human agency is not proportional to the effort required to be an agent.

If, in the far future, there are drugs or other enhancements that might enable athletes or sportsmen to score literally effortlessly without the necessity to add anything by way of skill, effort, or training, this would clearly abolish competition and any excitement it engenders. Whether this would be unethical or simply boring is another question of course. For my part I can see no place for the ethical to get a grip here except perhaps if the most severe versions of egalitarianism are right, in which case such enhancements would be mandatory for all. For, as W. S. Gilbert so eloquently pointed out:

> In short whoever you may be,
> To this conclusion you'll agree,
> When everybody's somebodee,
> Then no one's anybody![4]

Sandel recognizes the comparative weakness of the argument from effort so he digs deeper:[5]

> The deeper danger is that they (genetic enhancements) represent a kind of hyperagency—a Promethean aspiration to remake nature, including human nature, to serve our purposes and satisfy our desires. The problem is not the drift to mechanism but the drive to mastery. And what the drive to mastery misses and may even destroy is an appreciation of the gifted character of human powers and achievements.
>
> To acknowledge the giftedness of life is to recognize that our talents and powers are not wholly our own doing, despite the effort we expend to develop and to exercise them. It is also to recognize that not everything in the world is open to whatever use we may desire or devise. Appreciating the gifted quality of life constrains the Promethean project and conduces to a certain humility. It is in part a religious sensibility. But its resonance reaches beyond religion.

There are many problematic claims here. Why, for example, do we have to recognize and accept the gifted nature of normalcy but not the gifted nature of disease? An obvious answer is of course that it is rational to be choosy. But the choosy, if they are rational or fastidious, do not select on the basis of giftedness. So what is the basis for the differential acceptability of gifts? Sandel appeals to the problematic notion of honor and the complex notion of human "flourishing." He says, "Medical intervention to cure or prevent illness or restore the injured to health does not desecrate nature but honors it. Healing sickness or injury does not override a child's natural capacities but permits them to flourish." We will return to some of these issues later; for the moment we should record that Sandel's point cannot be that we must welcome the "gifted quality of life" only when it permits a child's (or an adult's?) natural capacities to flourish, because arguably enhancements do this when they enhance, improve upon, our "given" capacities. Do I flourish more or honor nature more if I protect myself against smallpox by vaccination or against cancer (we may hope) by a genetically engineered cell manipulation which confers immunity to cancer? Sandel is clearly issuing a general invitation to appreciate the gifted quality of life and, presumably, not to look gifted horses in the mouth even when those horses are a whiter shade of pale![6]

Surely we can recognize that "our talents and powers are not wholly our own doing, despite the effort we expend to develop and to exercise them" without committing ourselves to abjuring modifications which clearly are our own for the doing, and which would further enhance those (initially) unbidden powers and capacities.

When it is claimed that "not everything in the world is open to whatever use we may desire or devise," it is unclear what Sandel thinks he is saying. What is "not open" and what sort of locks are we talking about or what sort of closure are we looking for? The answer seems to be that the locks are powerful moral objections,[7] but what is the content of these objections? It seems to be the assertion that if there is one justified constraint, it is true that "not everything in the world is open to whatever use we may desire or devise," but nothing follows as to which things are not open in this way. Sandel seems to think the identification of the closed options is related to accepting or rather celebrating giftedness and "a certain humility." But why should we accept either of these? I personally do not regard humility as a virtue. Those who do have reason to find fault with my character, but the issue here is not the question of whether or not I am (or anyone is) less than perfect, or whether humility is a virtue or a vice,[8] but the question of whether or not I am to be free to be less humble than others and, if not more "gifted," at least cleverer or fitter if the technology is there to assist me.

Now, Sandel has a rather bizarre gloss on these ideas. He says:

> It is difficult to account for what we admire about human activity and achievement without drawing upon some version of this idea. Consider two types of athletic achievement. We appreciate players like Pete Rose, who are not blessed with great natural gifts but who manage, through striving,

> grit, and determination, to excel in their sport. But we also admire players like Joe DiMaggio, who display natural gifts with grace and effortlessness. Now, suppose we learned that both players took performance-enhancing drugs. Whose turn to drugs would we find more deeply disillusioning? Which aspect of the athletic ideal—effort or gift—would be more deeply offended?

That's a poser! I don't see any grounds for rational choice here, but Sandel knows the answer. Before looking at the answer we should note that there is also a somewhat spurious distinction employed here, since in most sports[9] "grafters" build on some natural talent and the ultra gifted usually also have to practice, train, and keep fit.

Some might say effort: the problem with drugs is that they provide a shortcut, a way to win without striving. But striving is not the point of sports; excellence is. And excellence consists at least partly in the display of natural talents and gifts that are no doing of the athlete who possesses them. This is an uncomfortable fact for democratic societies. We want to believe that success, in sports and in life, is something we earn, not something we inherit.

It is doubtful that drugs do provide a way to win without effort—athletes who use drugs must also train hard and have a high level of skill—but let that pass. It is odd that, having extolled the virtues of effort in an argument against enhancement, which admittedly Sandel admits is not decisive, Sandel then turns to praise of effortlessness.[10]

We do not, I believe, believe that "excellence consists at least partly in the display of natural talents and gifts." We may believe that sports should embody the display of natural gifts to some extent, but that is part of another argument. Excellence is being better, or rather best! The excellent in sports are the fastest, the strongest, the most skillful—those who score most goals, runs, home runs, points, or whatever—they are those who win, not necessarily those who win by displaying their natural gifts. Moreover, I don't think Sandel has a plausible notion of agency; excellence is not just something we are gifted, but rather it is something we do, something we are responsible for, something that is down to us. But as I have argued, this is the case when we make choices that have these effects. It is our choices for which we are responsible: not only our effortful choices but also our effortless ones, and not only our naturally gifted choices but our winning or successful choices that display other abilities. An interesting question is whether or not "luck" is a natural talent, a gift or even the result of wise choices. Napoleon is supposed to have prized "lucky generals"[11] but whether they were simply fortunate, or fortunate because brave or for other reasons, will never be known. What we can say is that luck is attributed to them on the basis of the decisions they made.

Sandel is just wrong to associate excellence with the natural; excellence has to do with excellence, with being very good, or being the best! Praise for excellence is appropriate when the agent is responsible for the excellence she achieves. He writes:

> The real problem with genetically altered athletes is that they corrupt athletic competition as a human activity that honors the cultivation and display of natural talents.[12]

Again it seems highly arbitrary to stipulate an account of authentic athletic competition which has as a necessary condition a highly artificial account of the natural.

Athletic competition is a human activity that honors the cultivation and display, not simply or even principally of natural talents, but of all talents, of all of those things for which we are responsible, either because they are ours (naturally) or ours because we have made them ours by our free choices.

> The ethic of giftedness, under siege in sports, persists in the practice of parenting. But here, too, bioengineering and genetic enhancement threaten to dislodge it. To appreciate children as gifts is to accept them as they come, not as objects of our design or products of our will or instruments of our ambition. Parental love is not contingent on the talents and attributes a child happens to have. We choose our friends and spouses at least partly on the basis of qualities we find attractive. But we do not choose our children. Their qualities are unpredictable, and even the most conscientious parents cannot be held wholly responsible for the kind of children they have. That is why parenthood, more than other human relationships, teaches what the theologian William F. May calls an "openness to the unbidden."[13]

Now, illnesses are unbidden, as are accidents, invasions by parasites and viruses and for that matter terrorists and foreign forces. I cannot see any obvious, or even subtle, merit in openness to the unbidden. Sandel has an answer to this too. He says:

> May's resonant phrase helps us see that the deepest moral objection to enhancement lies less in the perfection it seeks than in the human disposition it expresses and promotes. The problem is not that parents usurp the autonomy of a child they design. The problem lies in the hubris of the designing parents, in their drive to master the mystery of birth. Even if this disposition did not make parents tyrants to their children, it would disfigure the relation between parent and child, and deprive the parent of the humility and enlarged human sympathies that an openness to the unbidden can cultivate.

Openness to the unbidden is, according to Sandel, good when it is part of a nondisfiguring relationship between parent and child.

To appreciate children as gifts or blessings is not, of course, to be passive in the face of illness or disease. Medical intervention to cure or prevent illness, or restore the injured to health does not desecrate nature but honors it. Healing sickness or injury does not override a child's natural capacities but permits them to flourish.

Now we have to be open to the unbidden only when it is part of a nondisfiguring relationship between parent and child which does not desecrate nature but honors it. This is pure Humpty Dumpty! Sandel has piled up the resonant but opaque images until even Alice would feel at home in his wonderland, with words meaning just what you want them to mean.[14] While few would admit to deliberately desecrating nature, we are entitled to more by way of an account of why honoring nature in Sandel's sense is an obligation and why improving upon it artificially is a desecration, except when it is to cure or prevent illness. We need among many other things from Sandel a non-question-begging account of honor and desecration.

I have suggested that there is a continuum between treating dysfunction and enhancing function which invites us to consider the benevolent motives and life-enhancing outcomes of both.[15] I will not defend that again now, but for the moment we may simply note that Sandel stipulates a difference, gives resonant labels to the activity on each side of the distinction ("desecration" and "honoring"), but provides no arguments supporting the attachment of these labels or as to the meaning of the concepts as he employs them. Nor does he show how the activities they purport to describe merit such labels. He obviously regards it as self-evident, but alas, saying so does not make it so.

Sandel covers many other issues and has some interesting but inconclusive excursions into political philosophy and what he calls "bioethics," perhaps to distinguish that activity from his own way of philosophizing. He returns, at the last, to his former pairings of labels and, in fine rhetorical style, piles on a few more for good measure:

> But removing the coercion does not vindicate eugenics. The problem with eugenics and genetic engineering is that they represent the one-sided triumph of willfulness over giftedness, of dominion over reverence, of molding over beholding. Why, we may wonder, should we worry about this triumph? Why not shake off our unease about genetic enhancement as so much superstition? What would be lost if biotechnology dissolved our sense of giftedness?
>
> From a religious standpoint the answer is clear: To believe that our talents and powers are wholly our own doing is to misunderstand our place in creation, to confuse our role with God's. Religion is not the only source of reasons to care about giftedness, however. The moral stakes can also be described in secular terms. If bioengineering made the myth of the "self-made man" come true, it would be difficult to view our talents as gifts for which we are indebted, rather than as achievements for which we are responsible. This would transform three key features of our moral landscape: humility, responsibility, and solidarity.

It is true that an enhanced sense of human agency coupled with the reality of increased powers to influence the way the world is, and the ways it will be in the future, may transform our understanding of key features of our moral landscape. But to fail to recognize and respond to changes in the landscape is a recipe for disaster and it is difficult, perhaps impossible anymore, to view a world that we could change as anything other than a world for which we are responsible. When chasms open before us, only those open to the unbidden will walk straight on over the edge. The truth is that our understanding of the scope of our responsibility in and for the world including our responsibilities to and for our children has radically and rightly changed in recent times.[16]

Sandel is almost right when he says:

> As humility gives way, responsibility expands to daunting proportions. We attribute less to chance and more to choice. Parents become responsible for choosing, or failing to choose, the right traits for their children. Athletes become responsible for acquiring, or failing to acquire, the talents that will help their teams win.

But it is not failure of humility that has this effect, but a better understanding of responsibility in the presence of real and unavoidable choices.

One of the blessings of seeing ourselves as creatures of nature, God, or fortune is that we are not wholly responsible for the way we are.

This may be, but seeing ourselves in this way is an illusion; we may not be wholly responsible for the way we are, but our responsibility for the choices that have made us that way is inescapable:

> The more we become masters of our genetic endowments, the greater the burden we bear for the talents we have and the way we perform.

Again true, but inescapable: the burdens of responsibility are our fate as choosing autonomous beings. But mastery is the power to do things, not the exercise of that power. I am master of my fate if I *can* choose, not only if I *do* choose. I determine my destiny as much by declining or neglecting to exercise my mastery as by exercising it. This is a crucial point. Responsibility is a dimension of the ability to choose, not simply of a particular exercise of choice. The power of choice brings with it the burden of responsibility for the way we exercise choice, because choosing not to act is still a choice for which we are responsible. We cannot escape that burden by declining the *recognition* of our mastery of our fate or by choosing not to make decisions rather than making them in a particular way. Sandel seems to be recommending the use of *Joo Janta 200 Super Chromatic Peril Sensitive Sunglasses®* as patented by Douglas Adams in the second of his five-part Hitchhiker's Guide to the Galaxy Trilogy, *The Restaurant at the End of the Universe*.[17] These sunglasses are "specially designed to help people develop a relaxed attitude to danger. At the first hint of trouble they turn totally black and thus prevent you from seeing anything that might alarm you"[18] and thereby protect the wearer from the troubling necessity to take notice of (and hence to choose whether or not to act upon) any peril that threatens. The sunglasses may save the wearer from noticing the peril that she is in, but she is still responsible for wearing the glasses and for the consequences of so doing.[19] Sandel seems here to be recommending the equivalent. By refusing to use powers that we already possess, we somehow are supposed to be able to escape the responsibility of choosing (or not) to deploy them in particular circumstances. And what is true of individuals is true of societies; when a society denies individuals the opportunity of choices they could have, it takes upon itself that responsibility and in a democracy that responsibility is shared, it is collective.[20] A shared burden may be easier to bear but it is still a burden and the responsibility for that burden is not avoided.

The final point I wish to address concerns Sandel's linkage of moral obligation to the notion of giftedness:

> Why, after all, do the successful owe anything to the least-advantaged members of society? The best answer to this question leans heavily on the notion of giftedness. The natural talents that enable the successful to flourish are not their own doing but, rather, their good fortune—a result of the genetic lottery. If our genetic endowments are gifts, rather than achievements for which we can claim credit, it is a mistake and a conceit to assume that we are entitled to the full measure of the bounty they reap in a market economy. We therefore have an obligation to share this bounty with those who, through no fault of their own, lack comparable gifts.

Here Sandel is asking a big question and giving a small answer. No political theory I know of, nor indeed any theory of just taxation, requires us to share our gifts but keep our earnings. Indeed, income tax is predicated on the justice of sharing our earnings. It is true that governments sometimes levy so-called "windfall" taxes on unexpected "gifts" but these are rare and widely perceived as unjust. The best answers to the question of what the successful owe to the least advantaged (by, for example, Hobbes, Rousseau, Marx, Rawls, Sen, Dworkin) do not conspicuously rely on the notion of giftedness. If we have a moral obligation to prevent harm to our fellows or indeed to do them good (to enhance them), if we owe or feel solidarity or compassion or justice or charity or altruism to others, it is surely not because what we have to give to others to discharge this obligation cost us nothing in the first place—was a "gift" or came "unbidden" or unearned. This is mean-spiritedness taken to the extremes that only unbridled capitalism could tolerate.

More probably, however, Sandel is appealing here to so-called "luck-egalitarianism"[21], which holds that where people are worse off than others through no fault of their own they are therefore owed some form of compensation. However, if through my choices I ensure that my children are worse off than others either because I do not select against disability or disadvantage or because by failing to enhance them I leave them relatively deprived, I have created the very unfairness which Sandel says creates an "obligation to share" our "bounty with those who, through no fault of their own, lack comparable gifts." Surely it is more sensible to reduce the necessity to redistribute by creating individuals who are lucky rather than unlucky and therefore are more likely to possess a bounty that is shareable with others. Nothing in my own defense of enhancement begs the question against luck-egalitarians. Rather, enhancement provides more to redistribute and less need for redistribution. The more enhancement, the less bad luck and the more products of good fortune there are available for redistribution.[22]

This is Sandel's final flourish:

> There is something appealing, even intoxicating, about a vision of human freedom unfettered by the given. It may even be the case that the allure of that vision played a part in summoning the genomic age into being. It is often assumed that the powers of enhancement we now possess arose as an inadvertent by-product of biomedical progress—the genetic revolution came, so to speak, to cure disease,

> and stayed to tempt us with the prospect of enhancing our performance, designing our children, and perfecting our nature. That may have the story backwards. It is more plausible to view genetic engineering as the ultimate expression of our resolve to see ourselves astride the world, the masters of our nature. But that promise of mastery is flawed. It threatens to banish our appreciation of life as a gift, and to leave us with nothing to affirm or behold outside our own will.

This is of course an overstatement. But while the threats he sees are plausible, they are not compelling. We can equally affirm things within the choice, responsibility, and will of men. When we acknowledge that we are standing on the shoulders of giants we affirm what is within our will; when we acknowledge that we have an evolutionary inheritance upon which we can build we have something to affirm outside our own will. But suppose there comes a time when the given, unbidden inheritance is all but extinguished? What we then say will depend on our qualities of life and of mind. But these will not be better or worse in proportion to our responsibility for them anymore than the magnitude of a disaster is determined by whether or not it is natural or man-made, bidden or unbidden.

Humpty Dumpty, of whose approach Sandel's paper is so resonant, should have the last word:

> "When *I* use a word," Humpty Dumpty said, in a rather scornful tone, "it means just what I choose it to mean, neither more nor less."
>
> "The question is," said Alice, "whether you *can* make words mean so many different things."
>
> "The question is," said Humpty Dumpty, "which is to be master—that's all."[23]

And that really is the main issue I take with Sandels's poetic defense of the unbidden. If all that was at stake was Sandel's world view, or that of those who favor enhancements, there would be less need to take issue with his approach, with all its elegance and rhetoric. But Sandel is offering not only arguments to prefer his world view and reject another, but arguments to prefer his world view and *outlaw* another; he is preparing the ground for those who care "which is to be master." I am happy for Sandel to welcome the unbidden and accept all the gifts that come his way, including "the slings and arrows of outrageous fortune" which characteristically come out of the blue. But I wonder if he will let me and others choose the enhancements we prefer for ourselves and those which we judge best for our children? If he will, that is all well and good, but others will use his arguments to outlaw the option of enhancement. Those who wish to avail themselves of enhancements may wish to be master in one sense. They wish to be masters of their own destinies; I wish to be master of my destiny. But I, and I believe others who would wish to be free to choose enhancements, do not wish to be master in that other repressive, tyrannical sense of controlling the lives of others. Sandel can have the unbidden and welcome, but on condition he will let me and others have access to the bidden.

8 | Good and Bad Uses of Technology: Leon Kass and Jürgen Habermas

Four contemporary writers have been prominent in expressing particularly strong opposition to enhancement: Michael Sandel, Leon Kass, Jürgen Habermas, and Francis Fukuyama. Others already considered have also produced arguments which impact on the legitimacy of enhancement, but in a more nuanced way. We have already examined Sandel's arguments in some detail. In this chapter I will consider the case put by Kass and Habermas. Fukuyama I do not discuss in detail in this book, partly because his views have been examined effectively by Jonathan Glover[1] and partly because his arguments are for the most part duplicated by those discussed in detail here.

Leon Kass

Leon Kass has assayed a number of forays into the issues that surround enhancement. The most comprehensive account and perhaps the most expressive of his position is in an essay entitled "Ageless bodies, happy souls."[2]

Kass starts by being almost right about enhancements. He is right when he says:

> Needless arguments about whether or not something is or is not an "enhancement" get in the way of the proper question: What are the good and bad uses of biotechnological

power? What makes a use "good" or even just "acceptable"? It does not follow from the fact that a drug is being taken solely to satisfy one's desires that its use is objectionable. Conversely, certain inventions to restore what might seem to be natural functioning wholeness—for example, to enable postmenopausal women to bear children or 60-year-old men to keep playing professional ice-hockey—might well be dubious uses of biotechnological power.[3]

He then rehearses three obvious objections to enhancement that he believes fail. The first of these concerns safety, and Kass rightly comments that the "big issues have nothing to do with safety," for while safety is a big issue it is not a special problem for enhancement technologies or treatments, rather than for, say, nonenhancing therapies. Second, Kass considers the issues of fairness (stealing an unfair advantage) and distributive justice: the question of the fairness of some people being able to access advantageous technologies while others are not. We have discussed these issues in chapters 1 and 2 and again Kass[4] rightly says that "[t]he central matter is not equality of access, but the goodness or badness of the things being offered."

Finally, Kass considers the issues of freedom and coercion. Here Kass seems on less firm ground. He notes that "[e]ven partial control over genotype ... would add to existing social instruments of parental control and its risks of despotic rule. This is indeed one of the central arguments against human reproductive cloning: the charge of genetic despotism of one generation over the next." As I have argued elsewhere,[5] Kass is entirely wrong in this, there being no danger that human reproductive cloning would any more lead to despotism of one generation over another than does normal sexual reproduction. Sexual reproduction allows parents to determine a large part of the genome of potential offspring, and child rearing and education allow them the further license that the terms "care and control" of children imply. Since neither genotype nor parental wishes have significant impact on autonomy, these fears are somewhat exaggerated.[6]

Against this Kass notes that

[a]ttempts to alter our nature through biotechnology are different from both medicine and education or child-rearing. It

seems to me that we can more-or-less distinguish the pursuit of bodily and psychic perfection from the regular practice of medicine.... When it functions to restore from deviation some natural wholeness of the patient, medicine acts as a servant and aid to nature's own powers of self-healing. It is also questionable to conflate child-rearing and education of the young with the attitude that seeks willful control of our own nature. Parents do indeed shape their children, but usually with some at least tacit idea—most often informed by cultural teachings that have stood the test of time—of what it takes to grow up to live a decent, civilized, and independent life.[7]

Of course, most parents want their children to experience a decent, civilized, and independent life. But they also want willful control of the process, because they (probably) think they know best. Parents who would enhance their children think the same, informed by the same cultural teachings that have stood the test of time. Those teachings tell us to do the best for our kids and to give them whatever advantages we can.[8] The claim that attempts to alter our nature through biotechnology are different from both medicine and education or child-rearing seems wholly implausible. Medicine uses technology and biotechnology; indeed, much of medicine is a part of technology, it is a technological genre. For the rest, what matters surely is the ethics of altering our nature, not the means that we adopt. If it's right to alter our nature, we should choose the best and most reliable, not to mention the most efficient and economical, methods of so doing.

Kass offers nothing to replace the accounts of Boorse[9] and Daniels[10] critically discussed in chapter 3. There are distinctions in abundance, but distinctions with a moral difference are harder to find. No systematic account of relevant moral differences between biotechnological interventions and medical or social ones has been given by anyone and it seems doubtful that there could be such an account. There are, of course, moral differences between different sorts of interventions and these are the subject of the differences between Kass and myself. But these are not mirrored by the distinctions, such as they are, between therapy and enhancement nor between the normal and natural on the one hand and the artificial and unusual on the other. Recall that Kass himself has problems with this; he says that

> certain inventions to restore what might seem to be natural functioning wholeness—for example, to enable postmenopausal women to bear children or 60-year-old men to keep playing professional ice-hockey—might well be dubious uses of biotechnological power.

Kass obviously has strong intuitions that 60-year-old men should keep out of certain activities, but whether it is the ice that disagrees with them or with Kass, or the idea of their earning money, or whether it is something about the game of hockey that is particularly unsuitable beyond the sixtieth birthday, we are not told. This is not the place for a discussion of postmenopausal mums, but one wonders how older women and men have offended Kass.[11]

Cloning and Enhancement

Kass's deliberate linkage of cloning to enhancement may seem surprising, but he does so because he sees both as exercises in hubris, a presumptuous attempt to master nature and mold it to our liking. Kass's thoughts on cloning perhaps involve the idea that human clones will have psychological difficulties because people will have special expectations of them and possibly he imagines the same will be true of the expectations parents will have of their enhanced children. Just as the clone's life will always be compared with that of his or her genetic parent, the enhanced child will be compared with an imaginary ideal. It has been suggested that the clone risks having a "life in the shadow" of the person from whose genes he or she was cloned.[12] The idea is perhaps that a clone would have the feeling that his life had already been lived and, consequently, he will be deprived of "an open future." We know how his genetic parent lived, so we will know how the child will live. He will be considered as "the copy" that is of lower quality than its original and that has no life of his own. Elsewhere, Kass expressed it this way:

> The cloned individual, moreover, will be saddled with a genotype that has already lived.... People are likely always

> to compare his performances in life with that of his alter ego. Still, one must also expect parental and other efforts to shape this new life after the original—or at least to view the child with the original version always firmly in mind. Why else did they clone from the star basketball player, mathematician, and beauty queen—or even dear old dad—in the first place?[13]

Since there is no empirical data on human reproductive cloning, we have no idea from whom people would wish to clone, although it seems more likely that people will choose to clone themselves or loved ones than strangers or beauty queens. The most rational choice for a genome to clone would be that of a particularly healthy and long-lived individual. In such a case there seem to be strong arguments, both moral and prudential, to (if the technique were safe enough) avoid the risk of the genetic roulette that is sexual reproduction and opt for a tried and tested genome of proven virtue. Kass's argument assumes, without any evidence or plausible argument, the malevolence and stupidity of parents who want to make use of the cloning technique to have a child.

The second problem in appealing to the idea that a cloned child will not have an open future is saying that a clone will not be unique and will not develop a personal identity. We know that the clone (unless it is a monozygotic twin) will have a different uterine environment, will probably be born to a different woman at a different time and place to its genotype donor, will have a different education, and many other things. Since we also know that all these experiences actually affect the physical structure of the brain, there is no significant sense in which any clone could be determined to be like its genome donor.

The third problem with the argument is that all the concerns expressed in it are based on the assumption that people, despite all the explanations and information that exist to the contrary, will persist in their belief in genetic determinism, that whether the genome has been cloned or enhanced the resulting individual will have its freedom and capacity for self-development dictated by the genome it has been given. But we have all been created with a fixed genome without an obvious loss of freedom. Finally, such expectations are also true of all parenting. What parent does not look at their child with themselves in mind as a model of expectation, if only a model of minimal expectation? "I want my child to be better than me!" Even in a context of radical parental expectation it is unlikely that burdens on children will be so great as to render unacceptable the whole practice.[14]

Kass further believes there is a "special kind of restriction on freedom—let's call it the problem of conformity or homogenization [which] is in fact quite serious." Kass thinks the effect will involve "raising the floor but greatly lowering the ceiling of human possibility, and reducing the possibility of genuine freedom, individuality, and greatness."

These issues are addressed elsewhere in this book (in chapter 4), not least in the same context as Kass also readdresses them in relation to life extension. For the moment it is perhaps enough to observe that, just as eliminating a disease, say smallpox, contributes to conformity in the sense that people do not differ with respect to suffering or not suffering from smallpox, so also the conferral of new powers, of memory, concentration, strength, endurance, or intelligence will lead to a greater uniformity if not conformity (with its overtones of obedience) in the sense that people will possibly, albeit over a very long term, differ less drastically in respects amenable to enhancement. That is of course only if all are enhanced. If some are not then the differences will be greater between enhanced and nonenhanced individuals. This Kass should celebrate, but of course such exacerbation of difference also has its problems. However, Kass cannot have it both ways.

It is possible but not, I think, likely that people will differ less drastically in respects amenable to enhancement. It depends how the enhancements will work but, if they simply augment existing powers and capacities, then, since these differ from person to person, it is likely that the enhanced powers will also differ in this way and possibly exacerbate the differences.

Certainly enhancements will raise the floor, but surely also they will raise the ceiling of possibility for humankind. It may be that healthy people differ from one another less than all healthy people do from the sick, or that, as Tolstoy reminds us in the first lines of *Anna Karenina*: "All happy families are alike but an unhappy family is unhappy after its own fashion."[15] But these "facts" are hardly arguments for the preservation of disease or misery, although Kass is tempted by suspicion of happiness, as we shall see. It is worth recalling here our discussion of the evolution of education and literacy. People for sure have become almost uniformly literate in large parts of the world, but there is surely no evidence that this has led to any increase in conformity "reducing the possibility of genuine freedom, individuality, and greatness"—quite the contrary, surely?

Kass's Case against Enhancement

After his preambolic rehearsal of three of the less persuasive arguments against enhancement, Kass poses and attempts to answer the main question:

> Why, if at all, are we bothered by the voluntary *self*-administration of agents that would change our bodies or alter our minds? What is disquieting about our attempts to improve upon human nature, or even our own particular instance of it?[16]

Kass starts his answer cautiously:

> It is difficult to put this disquiet into words. We are in an area where initial repugnances are hard to translate into sound moral arguments. We are probably repelled by the idea of drugs that erase memories, or that change personalities; or of interventions that enable 70-year-olds to bear children or play professional sports.... But is there wisdom in this repugnance?

I feel Kass feels I ought to be ashamed to admit I am not repelled by any of these possibilities (though I would not necessarily choose them for myself). Kass has resumed his irrational flirtation with repugnance as a morally relevant feature; it is instructive to remind ourselves of its provenance and poverty.

In a now notorious discussion entitled "the wisdom of repugnance"[17] Kass tries to make plausible the thesis that thoughtlessness is a virtue. Below I draw on criticisms of this style of approach that I have elaborated elsewhere.[18] Kass begins again by inviting us to share his prejudices:

> We are repelled by the prospect of cloning human beings not because of the strangeness or novelty of the undertaking, but because we intuit and feel, immediately and

> without argument, the violation of things that we rightfully hold dear.[19]

Again I find myself failing to share Kass's relish for repugnance. Clones are surely as wonderful as any other humans and I am proud to say I am personally acquainted with quite a few clones. Moreover, human clones have existed as long as human reproduction has been practiced. Eve was in a sense "cloned" from Adam since the original meaning of the term derives from the Greek *Klon*, "twig" or "offshoot," and Adam's rib (so reliable[20] sources claim) played that role in the creation of Eve.[21] Monozygotic (identical) twins have also been around as long as sexual reproduction; they are clones and their occurrence is very frequent. Three per thousand births are clones in this (true) sense, sharing as they do their entire genome, unlike clones produced by cell nuclear transfer (the "Dolly method").[22] Cloning is therefore a familiar and largely successful reproductive practice, which occasions no repugnance in me and little in most others (with the possible exception of Kass himself).

The difficulty is, of course, to know when one's sense of outrage is evidence of something morally disturbing and when it is simply an expression of bare prejudice or something even more shameful. George Orwell[23] once memorably referred to this reliance on intuition as use of "moral nose"; as if one could simply sniff a situation and detect wickedness. The problem is that nasal reasoning is notoriously unreliable, and olfactory moral philosophy, despite valiant efforts by Kass and others, has done little to refine it or give it a respectable foundation. We should remember that in the recent past, among the many discreditable uses of so-called "moral feelings," people have been disgusted by the sight of Jews, black people, and, indeed, women being treated as equals and mixing on terms of equality with others. In the absence of convincing arguments, we should be suspicious of accepting the conclusions of those who use nasal reasoning as the basis of their moral convictions.

In Kass's suggestion (he disarmingly admits revulsion "is not an argument"), the giveaway is in his use of the term "rightfully." How can we know that revulsion, however sincerely or vividly felt, is occasioned by the violation of things we rightfully hold dear unless we have a theory, or at least an argument, about which of the things we happen to hold dear we *rightfully* hold dear? The term "rightfully" implies a judgment which confirms the respectability of the feelings. If it is simply one feeling confirming another, then we really are in the situation Wittgenstein lampooned as buying a second copy of the same newspaper to confirm the truth of what we read in the first.

In addition to the wisdom of repugnance, Kass has three main arguments to make. I shall follow the headings Kass himself uses for these arguments.

The Attitude of Mastery

In his discussion of mastery Kass follows Michael Sandel's arguments, which we have already examined in depth. Kass notes problems with them which have also been highlighted in this book. Noting Sandel's plea that we accept the gifted nature of existence, Kass rightly comments:

> Modesty born of gratitude for the world's "givenness" may enable us to recognize that not everything in the world is open to any use we may desire or devise, but it will not *by itself* teach us which things can be fiddled with and which should be left inviolate.[24]

To perform this trick Kass makes some distinctions:

> The word "given" has two relevant meanings, the second of which Sandel's account omits: "given" meaning "bestowed as a gift" and "given" ... something "granted" definitely fixed and specified. Most of the given bestowals of nature have their given species-specified *natures*: they are each and all of a given *sort*. ... To turn a man into a cockroach ... would be dehumanizing. To try to turn a man into more than a man might be so as well.[25]

Well so it might, but there is a big difference between the field of force of a word like "dehumanizing" and that of "wonderwoman" or "superman." The issue, just as Kass rightly points out when discussing Sandel's account of the given, is not whether it is *given*, nor even whether it is "species-specified" or "natural," but whether it is *good* to turn a man into a nonhuman person.

Kass accordingly makes the point that

> [o]nly if there is something precious in the given—beyond the mere fact of its giftedness—does what is given serve as a source of restraint against efforts that would degrade it.

There follows a long and wistful list of "only if's":

> Only if there is something inherently good or dignified about, say, natural procreation; ... only then can we begin to see why those aspects of our nature need to be defended.

Unfortunately all these "only if's" involve an equally long list of non sequiturs when we come to the "only then's". Those who have tried it often report that there is not much dignity involved in natural procreation, either at the stage of conception or indeed at birth[26]. In the case of conception, the lack of dignity is often cited as part of the fun: "if God did not have a sense of humor, he / she wouldn't have invented sex." Be that as it may, it does not follow from the fact that there is something good or dignified about a natural process that it needs to be defended for the simple and sufficient reason that there may be something better about its synthetic modification of replacement. It is whether or not *that* is the case which would need to be established before the inherent goodness of something could function as a reason for defending it.

"Unnatural" Means

When we turn to the nature of the means, Kass identifies the question as

> how do, and how should, the excellent ones become excellent?

He is concerned with a whole range of what he regards as dehumanizing means of achieving human enhancement, methods which deform the "deep structure of natural human activity."[27] For example, when he says "'personal achievements' impersonally achieved are not truly the achievements of persons. That I can use a calculator to do my arithmetic does not make me a knower of arithmetic," he is surely right, but we do not condemn calculators nor indeed their users on that account; rather, we use them without particular pride, although knowing when to use a calculator is something one can be proud of and knowing when to choose enhancement for oneself or others and which enhancements to choose may be achievements of considerable significance.

Kass believes that the fact that we are not responsible for the effects of chemical or electronic enhancers, and hence cannot take proper pride in the achievements they facilitate, somehow makes their use inappropriate or inauthentic.

This may be true of some types of enhancement, but for all of them we may surely take pride in our choice of appropriate means to our ends and congratulate ourselves on our wise choices and on the fact that we have made the choice that benefits us in ways that we value. Some chemical, mechanical, or electronic enhancers will leave plenty of room for effort, skill training, hard work, suffering, and all the other good things particularly favored by puritans. If I take a pill which improves my memory or powers of concentration, I will still have to study, and use that study to draw conclusions, formulate ideas, or write books for which I may take appropriate credit or blame: credit for the work and ideas but not the powers of concentration or memory that helped me to those achievements. Equally, if I take another pill which improves my stamina and endurance, or my muscles are enhanced by drugs, I will still have to train and I will probably also still need some natural (whatever that may be) talent for the sport or athletic event in which I wish to participate. There is always some personal contribution, perhaps unhappily for those true consequentialists who just value the consequences, however achieved, or those men and women who just value their superiority, however effortless.[28]

Kass ends this section with many words betraying the promise of one:

> In a word, one major trouble with biotechnical (especially mental) "improvers" is that they produce changes in us by disrupting the normal character of human being-at-work-in-the-world…which when fine and full constitutes human flourishing.

One feels inclined here to say "well, whatever turns you on Leon," but there does not seem to be a compelling value engaged here. *Au contraire*, many favor "human being-at-rest-in-the-world" or human being-having-a-good-time-in-the-world." Kass concludes this section with the observation that "[t]o the extent that we come to regard our transformed nature as normal, we shall have forgotten what we lost." Perhaps, but consider a phenomenon of two centuries ago: "human being-traveling-to-New York-from-London-in-the-world-and-taking-three-months-to-get-there" or "human being-without-anesthesia-and-suffering-the-pains-of-natural-childbirth." True, we have forgotten what we lost; the issue is, as so often, one of rational regret. Do we have any reasons to regret this loss when we compare the loss with our present situation? Remember those who wish to can still spend months at sea without vitamin C and with scurvy a constant companion; and those who wish to can have fully natural childbirth, however painful or risky. What is different is that it chances that these things are now matters of choice.[29] Indeed, the fact that they are matters of choice, and yet still people sometimes choose the hard way, is further evidence that Kass's Luddite approach to enhancement is misplaced, even for someone with his priorities. People still choose to sail round the world, or trek across the Arctic with dog sleighs, and some benighted souls even refuse anesthesia. Technology, including enhancement technology, surely gives us the best of both worlds, the Luddite and the real.

Dubious Ends

The endgame is, not untypically, dubious. In this, the final section of Kass's essay, he concentrates on longevity and his arguments against it. Since I have considered these in chapter 4, I will not discuss them here. Kass next considers "pharmacologically assisted happy souls."[30] He notes that "[e]rsatz ... feelings of self-esteem are not the real McCoy." Forgive a philosophical quibble, but this remark is analytic: it is necessarily true and tells us nothing. Of course "false" feelings are not real feelings! Kass elaborates:

> No music lover would be satisfied with getting from a pill the pleasure of listening to Mozart without ever hearing the music. Most people want both to feel good and to feel good about themselves, but only as a result of being good and doing good.[31]

Again Kass is right, but not because he or the music lover to whom he appeals are fastidious. You cannot get music from a pill because a pill does not make a sound, even in the brain. In a related way, you cannot have a pill to make you feel "grief" because grief is not a bodily sensation. Grief is the name we give to the feelings (of whatever sort) occasioned by loss through death. No death[32], no grief; that again is analytic, true by definition. It is not a failure of technology that there is not and could never be a pill for grief. It is a logical not a technological problem. Despite this error, Kass is still right: we don't want a pill to make us feel that our lover loves us—we want to be loved. Equally, we don't want a pill to make us feel rich—we want to *be* rich. So we won't use pills for love or money. But if we are not rational we might try, and serve us right! But what's that to the purpose? It is not wrong, nor unethical, merely (merely!) futile.

There is a sense of course in which we do have a technological source of Mozart on tap. It is recorded music, from phonograph to iPod. But the use of these has not supplanted live music: concerts, whether rocky-horrific or chamber Mozartian, are still alive and well. An additional fallacy of Kass's approach is that he assumes, contrary to the evidence, that new technologies always supplant old methods or natural processes. Usually they coexist; people are happy to use their iPods on a daily basis but they still go to concerts and many still play instruments. My eighteen-year-old son loves his iPod, he plays the flute and goes to concerts, and many others do equivalent things.

We might, however, want a pill to make us concentrate, stay awake, or improve memory, but since we cannot remember what we haven't experienced it will be no substitute for learning, whether by rote or by experience. Just as the straw men Kass has invented are not real men with real arguments, so his ersatz experiences are not real experiences: not experiences at all because not experienced, merely felt.

This was all discussed long ago by Jonathan Glover,[33] who imagined an experience machine that might make us believe we owned, say, expensive beachfront property in California and were experiencing the attentions of many beautiful lovers of infinite attractiveness and skill. Why would we not "live" on the experience machine forever? Why would it not be chosen by rational self-interested people. In fact, Glover and I would both, with Kass, reject that machine for the reasons set out so eloquently by Glover and Kass. But this is also because those reasons mean that the machine would constitute no sort of enhancement of life or of people. Is it unethical? Should we prevent others from accessing it if they could pay for their infinite experience? A different question!

Kass's final flourish is full of sound and fury but it signifies only a puritan mean-spiritedness and lack of imagination:

> A flourishing human life is not a life lived with an ageless body or untroubled soul, but rather a life lived in rhythmed time, mindful of time's limits, appreciative of each season and filled first of all with those intimate human relations that are ours only because we are born, age, replace ourselves, decline and die—and know it. It is a life of aspiration, made possible by and born of experienced lack, of the disproportion between the transcendent longings of the soul and the limited capacities of our bodies and minds. It is a life that stretches towards some fulfilment to which our natural human soul has been oriented, and, unless we extirpate the source, will always be oriented. It is a life not of better genes and enhancing chemicals but of love and friendship, song and dance, speech and deed, working and learning, revering and worshipping. The pursuit of an ageless body and a satisfied soul is finally a distraction and a deformation.

It is just magic dragonalia to think that all the good things in this list are only possible because of limited time, limited imagination, and death. As Tom Stoppard said in another context: "necrophiliac rubbish!"[34] The deformation is not wishing for and trying to achieve something different and better, but in trying to prevent those with a different vision from even attempting to live without dissatisfaction and death. I wish for the possibility of removing limits from our bodies and minds. I believe this may become fully possible without abandoning a life of indefinite extent full "of love and friendship, song and dance, speech and deed, working and learning" and of course much fuller of these wonderful things than Kass's stunted lives could ever be. I personally am not much attracted to "revering and worshipping"; to me these dispositions are entirely devoid of dignity, human or personal, but Kass may indulge in them, and welcome. What is thoroughly unacceptable to any free spirit is to accept Kass's puritan, narrow, limited, and undignified, let alone weary, stale, flat, and unprofitable, conception of the scope and limits of life and its possibilities. May his soul have its life's desire and forever remain unsatisfied. He is truly welcome. But is it not a little tyrannical to purport to stand in the way of the very different aspirations of others?

Kass's final flourish sets out a self-defeating and defeatist agenda. He aspires to finitude, to preset and arbitrary limits to human aspirations as well as to human powers:

> Finitude recognized spurs aspiration. Fine aspiration acted upon *is itself* the core of happiness. Not the agelessness of the body, nor the contentment of the soul, nor even the list of external achievement and accomplishments of life, but the engaged and energetic being-at-work of what nature uniquely gave us is what we need to treasure and defend.[35]

I do not recognize finitude, only the limitless possibilities of the human spirit and of human ingenuity. If Kass wishes to remain a limited human being, that is of course a matter for him. The sinister message is in the tail of Kass's agenda; he wishes to "treasure and defend." Treasuring is fine—Kass himself is something of an international treasure—but when he seeks to defend limitations he trespasses on the freedom of others and seeks to shackle the human spirit within the confines of the limits of his own desires and imagination. This is something those who defend freedom will wish to oppose.

Jürgen Habermas

In 2003 Jürgen Habermas published a book entitled *The Future of Human Nature*, making available in English translation ideas of his which were first presented in 2000 and 2001. This book, setting out his opposition to the human determination of human nature, is excruciatingly complex and crushingly conservative.[36] The stature of Habermas as one of the most famous and influential German philosophers of the last century makes it an important work to understand. The book discusses many issues of contemporary relevance from human cloning to sex selection, stem cell research, and preimplantation diagnosis. Two sets of arguments are particularly relevant to human enhancement and it is these on which we will concentrate now.

We will start with Habermas's ideas concerning the evils of actions which might control or preempt the decisions of future generations.

Future Bondage of the Living to the Dead

Habermas takes over ideas he attributes to Hans Jonas. In a resonant passage, Jonas sets out his stall against the legitimacy of decisions which might determine the nature of future generations:[37]

> But whose power is this—and over whom or over what? Obviously the power of those living today over those coming after them, who will be the defenseless objects of prior choices made by the planners of today. The other side of the power of today is the future bondage of the living to the dead.[38]

Habermas himself repeats and endorses this idea of the illegitimacy of eugenic control many times in the course of the book. He seems to have two main objections. One is in essence that voiced by Jonas, that of human bondage. The second stipulates, but produces only the most obscurantist of arguments for, the idea that such determination is inegalitarian and destroys in members of subsequent generations "an equal right to an autonomous conduct of life."[39]

How does Habermas develop these ideas?

> With genetic enhancement, there is no communicative scope for the projected child to be addressed as a second person and to be involved in a communication process. From the adolescent's perspective, an instrumental determination cannot, like a pathogenic socialization process, be revised by "critical re-appraisal." It does not permit the adolescent looking back on the prenatal intervention to engage in a *revisionary learning process*.[40]

> Eugenic interventions aiming at enhancement reduce ethical freedom insofar as they tie down the person concerned

> to rejected, but irreversible intentions of third parties, barring him from the spontaneous self-perception of being the undivided author of his own life.[41]

Irreversibility and Lack of Consent

We noted in chapter 5 that if consent were required before we could do things for or to children, few children would survive long enough to grow to adulthood and the consequent cruelty to children would reach unprecedented proportions.

It is surely obvious that we cannot avoid making decisions when failure to do so may adversely affect others, not only children but also for future generations. To decide not to intervene to enhance where we can do so is to condemn future generations to life without the advantages we might have bestowed. They can no more consent to this deprivation than they can consent to the interventions to which Habermas objects:

> [T]here is no communicative scope for the projected child to be addressed as a second person and to be involved in a communication process. From the adolescent's perspective, an instrumental determination cannot, like a pathogenic socialization process, be revised by "critical re-appraisal." It does not permit the adolescent looking back on the prenatal intervention to engage in a *revisionary* learning process.

Of course not, but the same is true for the instrumental determination involved in decisions not to protect or advantage children. A child cannot do any of this long-winded jargonized thinking with a "third party" decision to intervene, nor can he or she do so with decisions not to intervene. Consider the decision to vaccinate shortly after birth or a decision not to take folic acid before conception: the "future bondage of the living to the dead" is a permanent feature of generational life. We have noted elsewhere in this book the ways in which the sights and sounds to which our parents expose us irrevocably create the connections in our brains and hence the functioning of our minds. It has not been threatened by the possibility of human enhancement any more than the myriad of other prior decisions that have determined the nature of the world we have inherited and the bodies and minds we possess.

Ethical Freedom and Equality

Habermas's suggestion is that because future (enhanced) generations "may no longer see themselves as the undivided authors of their life," they may also find themselves with interpersonal relationships that are "no longer consistent with the egalitarian premises of morality and law."[42] Why this should be is, if possible, even more obscure, but Habermas seems to rely on the following as support for this idea:

> In the context of democratically constituted pluralistic society where every citizen has an equal right to an autonomous conduct of life, practices of enhancing eugenics cannot be "normalized" in a legitimate way, because the selection of desirable dispositions cannot be *a priori* dissociated from the judgment of specific life projects.[43]

I think Habermas is saying that because future enhanced individuals will not have chosen their enhanced nature they cannot conduct their lives autonomously. We have seen, however, that we are all in the position of having had "the way we are" determined by a combination of the acts and omissions of our parents and others with whom we have interacted since conception. If this is inimical to equality or autonomy, then neither equality not autonomy exist nor have they ever existed.

There is a story told of the former British prime minister James Callaghan. When he was chancellor of the exchequer he thought he should brush up his skills in economics. He apparently received private tuition at Nuffield College, Oxford. A student who shared the same economics tutor as the future prime minister asked his tutor what Callaghan was like as a student. The answer was: "James Callaghan is a very hard man to bamboozle, but once bamboozled, a very hard man to unbamboozle!"[44] The best explanation I can find for Habermas's views on "critical reappraisal" and "*revisionary* learning" is that he has managed to convince himself that these ideas have meaning and relevance in the context of enhancement and has, like Callaghan, proved very hard to unbamboozle.

Habermas has another go at these issues in his "Postscript 2002."[45] He suggests that if we want to understand the harms of enhancement correctly we need to consider them in the light of a model of a particular moral community:

> According to this model, eugenic practices, while they are not *directly* intervening into the genetically modified person's spheres of free action, might well harm the status of the future person as a member of the universe of moral beings.... In the moral universe, subjection of a person to the unjustly imposed arbitrary will of another one is ruled out.

Habermas then contrasts this with:

> An external or alien determination of the natural and mental constitution of a *future* person, prior to an entry into the moral community.

Habermas seems to regard this both as a serious possibility and as being perhaps even worse than slavery. Although when he implies that either "might well" harm the status of the future person, he is saying something pathetically weak. It might well not! There is no way that any of the methods of enhancement considered in this book or known to science could have the effect of determining "the natural and mental constitution of a *future* person" if this implies some loss or autonomy. This is scaremongering of an order that could surely only be the product of a bamboozled mind. If bamboozling of this depth and comprehensiveness might be the product of enhancement, then indeed there would be good reasons for apprehension for the future of humankind!

Consider the example of such an alien intervention with rather different consequences that we discussed in chapter 2. We considered David Baltimore's work to develop genetic interventions to prevent cancer and heart disease in future generations. If these were developed, then even if (per impossibile as I think) it might harm the status of a future person as a member of the universe of moral beings, it would be a great advantage to her as a member of the universe of healthy beings.

But is there any reason to suppose, as Habermas hypothesizes, that genetic enhancements threaten membership of the universe of moral beings? It seems to come down to how Habermas imagines that the future people will feel about knowing they have benefited from enhancements. Habermas sees only gloom and doom:

> Insofar as the genetically altered person feels that the scope for a possible use of her ethical freedom has been intentionally changed by a prenatal design she may suffer from the consciousness of sharing authorship of her own life and her own destiny with someone else.[46]

The "insofar as" may be no distance at all and the "may" may be not. One feels that this possible person should pull herself together and be glad that her parents cared enough about her to choose to give her advantages and protections, the lack of which might have harmed her more! Since we cannot by hypothesis know either way, we cannot be sure. We are necessarily judging now sub specie aeterni[47] and her parents, trying to do their best, have a judgment to make. That judgment and the responsibilities that go with it cannot be avoided by the ostrich-ethics recommended by Habermas.

But it gets worse. Habermas has a final doom-laden flourish, but this time the doom is sure and certain, not possible or maybe:

> Not even the highly general good of bodily health maintains one and the same value within the contexts of different life histories. Parents can't even know whether a mild physical handicap may not prove in the end to be an advantage for their child.[48]

Of course they can't! And they equally cannot know whether a severe handicap might have this effect. So what are we to conclude: that we should consider handicapping our kids just in case? Or what is the same thing,[49] consider, and actually decide, not to remove possible handicaps just in case?

Parents cannot evade or avoid responsibility for how their children will turn out, at least insofar as they have the power to leave things as they are or make them different. The power to make things different means that parents have a choice to do or forebear with highly probable but different consequences attending either future possibility following their decision. They have to do their best. What they should not do is their worst for their kids, and leaving their kids with a handicap of unknown and highly speculative advantage but known disadvantage would be, to put it mildly, problematic in almost anyone's philosophy.

9 | Designer Children

The phrase "designer children" is almost always used pejoratively. But if the design is an improvement on the random combination of genes or if it is more reliable, could it be ethical *not* to be a designer? Even if we are right to disapprove of the motives of many parents who seek to influence how children will turn out, it is far from clear that we are entitled to formalize our disapproval in prohibitive legislation or regulation. We examined some of these problems in chapter 5. Here we will take a different focus and look in some detail at an element of design—gender—which is in no sense an enhancement. Looking in detail at the ethics of influencing the nature of future people when there is no question that the altered nature will constitute an improvement will tell us much about the ethics of determining what sorts of people there will be and indeed of intervening in evolution.

We need to consider both clear cases of interventions that are not enhancing at all and marginal cases where the question of whether or not the intervention improved things is not susceptible of a clear answer. One reason for doing so is that if such neutral or problematic interventions are both ethical and defensible then, a fortiori, those which bring palpable benefits will be too. We will start with marginal cases and move on to a clear case of desired interventions which do not confer anything by way of enhancement or evolutionary advantage.

Marginal Cases

There will always be cases that some people consider enhancements and others would not. In Victorian times the virtues of patriotism and piety were much in vogue.[1] Today they seem less attractive to many. What if someone today wanted to endow their children with enhanced patriotism and piety? It is doubtful if we would consider (even if we believed in "the two P's") that people would be much harmed by having only the normal quota of these feelings so it is doubtful if anyone would regard such modifications as a matter of obligation along lines developed in previous chapters. However, should people be permitted to make such modifications if they proved possible?

Perhaps it is fortunate that modifications to such nebulous emotions (or are they dispositions?) as patriotism or piety are very unlikely to be possible, not least because such feelings are also likely to be correlated with other emotions which even those who value "the two P's" may feel uncomfortable about. But suppose such modifications were possible. Or suppose that there was indeed a gene for homosexuality and some parents felt that protection from, or indeed guarantee of, a disposition to homosexuality would be a distinct enhancement. There are characteristics which are what I would call "morally neutral"; gender seems a clear case since it is not morally speaking "better" to be female rather than male or vice versa. However, gender is perhaps an identity-determining feature rather than a phenotypical difference which a particular person might or might not have (say, eye color or hair color). The issue of identity-determining features is important because if the feature in question determines identity, then it is not something inflicted on or denied to a particular person. Rather, we are talking about the ethics of creating one person rather than another. In this latter case it seems that so long as the feature is neutral (neither better not worse to be) then if it is not wrong to be such a person, not harmful, undignified, disadvantaged in any serious way, then it cannot be bad or wrong to create a person with those features. The issue is rather different if we are imposing a particular condition on an individual who would otherwise exist without that feature.

Let us now turn to a clearly neutral exercise of the power to determine how (and perhaps who) future people will be.

Playing Fairy Godmother

Is it morally wrong to wish and hope for a fine baby girl or boy? Is it wrong to wish and hope that one's child will not be born disabled? I assume that my feeling that such hopes and wishes are not wrong is shared by every sane decent person. Now consider whether it would be wrong to wish and hope for the reverse? What would we think of someone who hoped and wished that their child would be born with disability, impairment, or even some disadvantage that fell short of whatever high standards are required before we classify a condition as a disability or impairment? What would we think of someone who wishes even a mild illness or injury for their child? Again I need not spell out the answer to these questions.

But now let's bridge the gap between thought and action, between hopes and wishes and their fulfillment. What would we think of someone who, hoping and wishing for a fine healthy child, declined to take the steps necessary to secure this outcome when such steps were open to them? Again I assume that unless those steps could be shown to be morally unacceptable our conclusions would be the same.

We have examined these questions extensively so far and have found that the answer to them is an emphatic "yes," we should make such improvements or enhancements if we can. However, many people think that elements of "design" which are not obviously health related or clearly beneficial are not only not morally required but morally bankrupt in a way that not only licenses disapprobation but licenses legal measures to ensure that we protect our children and our society from them.

Let us then re-pose our question: If it's not wrong to wish for a bonny, bouncing, brown-eyed baby girl, why or how would it become wrong if we had the technology, the choice, to play fairy godmother to ourselves and grant our own wishes?[2]

We should remember that the traditional way of producing children, namely by selecting a marriage or, less formally, a procreational partner, is very often governed by prejudices or preferences, not only for a particular sort of partner, but for the particular sort of child that mating with that partner will produce.

Gender Selection

If we examine a very basic element of design, namely gender, as our exemplar of an attempt to produce children to a particular "design," we can become much clearer about the ethics of all such interventions. Gender selection has been one of the most controversial areas of possible parental choice. An investigation of the ethics of gender selection will, however, have consequences for other morally neutral traits like hair and eye color, physique, and so on. I say "morally neutral" because I assume no reasonable person thinks it could be *morally* better to have one color of hair rather than another, nor for that matter to be one gender rather than another. Although this is often taken to be a difficult question and indeed the idea of parents being able to choose such things very often causes outrage, it seems to me to come to this: either such traits as hair color, eye color, gender, and the like are important or they are not. If they are *not* important why not let people choose? And if they *are* important, can it be right to leave such important matters to chance?[3]

Of course the *manner of choosing* may be morally important. We might feel that abortion was not a reasonable way of determining such things; but if, for example, a liter of orange juice taken at a particular point in pregnancy could achieve the desired outcome, I doubt if any attempt to regulate its use would succeed or would even be contemplated.

Objections to the idea of gender selection and the like often turn on two forms of "slippery slope" argument. Either it is claimed that a pattern of gender preference will emerge which will constitute a sort of "slap in the face" to the gender discriminated against, an insult and humiliation, like a piece of racist graffiti perhaps, or it is suggested that the pattern of preference will be such as to create severe imbalance in the population of society with harmful social consequences. Plainly these are very different sorts of outcome.

We should note that a pattern of preference for one gender amongst those opting for gender selection would not necessarily be evidence of sexist discrimination. There might be all sorts of respectable, non-prejudicial reasons for preferring one gender to another including just having a preference for sons or daughters. A preference for producing a child of a particular gender no more necessarily implies discrimination against members of the alternate gender than does choosing to marry a co-religionist, a compatriot, or someone of the same race or even class implies discrimination against other religions, nations, races, or classes. Of course, if a pattern of preference in favor of one gender were to emerge, it might have either or both of the effects we have noted and would certainly be cause for concern. However, it seems to be verging on hysteria simply to assume either that it would inevitably have these effects or that the effects would be so damaging as to warrant legislation to prevent the remotest risk of their occurring.

Some might claim that such choices might be insulting in the way perhaps that graffiti of a swastika[4] drawn by a Buddhist as a representation of a traditional Buddhist symbol might nevertheless be offensive to Holocaust victims, thus creating a defeasible reason not to do it. Of course we always have reasons not to offend others but both the graffiti example and gender selection illustrate clearly the power of good reasons for doing things that might offend others (in these cases freedom of speech and religion and reproductive autonomy) trump the sensibilities of those susceptible to offense.

Gender Is a Morally Neutral Trait

It is not ethically speaking better to be a boy rather than to be a girl. To say that a human trait or feature is morally neutral is to say that there is no moral reason to prefer to have that trait nor to be without it, no moral reason to try to create that trait or feature, nor any reason to try to eliminate it, no moral reason to hope for it or fear it. In short, gender is not normally the occasion for *rational regret*. Of course some few individuals who believe that they have been born with the wrong physical manifestations of gender do regret having these and often request gender reassignment surgery. I note this phenomenon, but to investigate it further is beyond our present concerns.[5]

Absent moral reasons for preference, there is a presumption in favor of free choice. This is of course a rebuttable presumption, but the reasons for so doing must be powerful. I will return to this point in a moment.

The question then about determining sexual orientation or indeed degrees of patriotism or piety, in the unlikely event that determining such things should ever become possible, becomes one of the reasonableness of deliberately creating a person who will have, but might not have had, these characteristics. This may have to be judged on a case-by-case basis but perhaps the issue of gender selection just discussed is some guide.

If we ask what level of risk a society like the United Kingdom, for example, should be prepared to run in order to avoid the unnecessary imposition of legal restraint on choice and to see whether a pattern of preference emerges and if so whether or not it is a pattern of discriminatory preferences, the answer is not clear.

A Modest Proposal for a Licensing Scheme[6]

My own proposal would be that a society like the United Kingdom, say, of about fifty-eight million people, could afford to license, say, one million procedures for gender selection over a ten-year period with options to revise the policy if severe imbalance seemed likely and was likely to prove significantly damaging to individuals or society. The United States could of course afford to license maybe six times as many procedures. With a licensing scheme we could then see what patterns of selection and motivation emerged. Even if all choices went one way, the imbalance created would be relatively small before detection, and a halt could be called if this seemed justifiable. I doubt that the places allocated on such a program would not be taken up (it would of course be self-financing and would not be part of the public health care system). It must be remembered that those who opted for gender selection would (with current technology) have to be very circumspect about their procreation and use sperm selection or preimplantation testing as the method. This would not, I guess, be wildly attractive or indeed particularly reliable. For the foreseeable future the take-up will also by limited by the availability of clinics offering the service. In any event, the way forward for a tolerant society respectful of autonomy would surely be not to rush to prohibitive legislation, but rather to license the activity with regular monitoring and see whether anything so terrible that it required prohibitive legislation emerged.

Suppose that an unprecedented perturbation of the spheres, or other natural event, had caused a one-off increase in births of one gender, say by 1%? Would the prospect of more boys or girls (even a million or so) than would otherwise have been expected throw U.K. society into a spin? Would six million more Americans of a particular gender cause panic? Would anyone lose any sleep over it? Suppose we had to wake up to the fact that there *already was* a gender imbalance in our different societies in favor of girls? Suppose (perish the thought) that in the United Kingdom there were already about 1,250,000 more women than men! In such a terrible eventuality perhaps we should hope that gender selection would predominantly favor males, and that some inroads into this adverse balance might thereby be made. Of course this is in fact the situation we currently face in the United Kingdom with just that (favorable?) gender balance in favor of females.[7]

A Case Study: Gender Selection

A report by the United Kingdom Human Fertilisation and Embryology Authority (HFEA)[8] recommended outlawing gender selection[9] and it is instructive as well as interesting to examine their reasoning, which is typical of much international opposition to gender selection.

The HFEA had conducted a consultation exercise prior to publishing its report, canvassing opinion about the legitimacy of gender selection, and the results of this wide public consultation were clearly influential.

A famous essay on Jane Austen by D. W. Harding[10] argues that "gentle Jane's" novels of English country life are far from gentle and are best understood as subtly expressing Austen's hatred for the small-minded and petty bourgeois world, full of the prejudices and conceits she describes: a hatred regulated by her ironic and deft prose. The HFEA report embodies a sort of mirror image of Harding's insight. In the report, the opinions which emerge from the consultation exercise, which, unsupported by evidence or valid arguments, are impossible to distinguish from prejudices, are given formal approval and proposed regulation by a government-appointed body set up with the responsibility to provide expert leadership. The leadership in fact exemplified in this report owes much to the legendary Duke of Plaza Toro, so faithfully and tellingly recounted by W. S. Gilbert.[11]

In her introduction to the HFEA's report, Suzi Leather, then Chair of the HFEA, remarked:

> I consider that our conclusions and the advice contained in this report represent an informed, balanced and proportionate response to the very complex issues raised by sex selection and I hope it will stand as a principal point of reference for all those—Government, professionals and the interested public—who will be involved in taking the debate forward.

While undoubtedly informed, it is informed largely by the results of a public consultation, the "hostility" of which to gender selection is manifest and even explicitly acknowledged in those terms by the HFEA, and this apparent hostility is accepted at face value. But more importantly the HFEA's report is inconsistent and, in the rare cases in which arguments appear, the arguments do not adequately support the conclusions drawn. The form of the document is also revealing. The bulk of it, i.e., chapters 1–4 (of six chapters), concerns background, and gives an overview of the reportage of commissioned research and public attitudes. Virtually the whole of the burden of establishing the report's conclusions is permitted to, and clearly does, emerge from the public consultation. Whether or not the public consultation was informative of the debate, it was in effect determinative of the conclusions that the HFEA reached. After these first four chapters, the report moves swiftly to its conclusions without any demonstration of balancing or considering the relevant arguments. These are manifest defects but they are not untypical of many national and international documents on gender selection and other elements of design. The reason (or excuse) for considering this report and its defects in some (albeit brief) detail is their representative nature.

An Examination of the Report's Conclusions

The HFEA came out strongly against all but strictly therapeutic uses of gender selection; and by "therapeutic" is meant uses which prevent the passing on of sex-linked disorders. Let us start with the inconsistency involved in this exception. A first thing to note is that the HFEA were cautious. They ruled out one method of gender selection, namely flow cytometry, because "[i]t is not possible to discount a theoretical risk to health with the use of this technique." Of course one can never discount risks if this absurdly high standard of caution is employed. Such a standard would rule out the benefits of almost all medical procedures, because there is always a theoretical risk, if not an actual risk, however slight. It may be that this method of gender selection should not be used on safety grounds, but not because there is "a theoretical risk to health," but surely because there is a real and significant risk. The same of course goes for many objections to new technologies, including the enhancement technologies considered in this book.

The HFEA then go on to consider so-called "gradient methods" of sex selection and note that "there is no reason to suspect that gradient methods pose a significant risk to the health of offspring." We should note in passing, of course, that even this certificate of risk-free health given by the HFEA would fail their previously employed rigorous test, namely that "it is not possible to discount a theoretical risk to health." Since it is never possible to discount such a risk, if this is an objection it applies to gradient methods as powerfully as it does to flow cytometry. However, let that pass. What is clear is that the HFEA regard even minimal risks to the health of resulting children which may flow from risks inherent in the methods of sex selection (rather than any social or psychological outcomes) as decisive in rejecting gender selection except where it would be used to rule out the inheritance of sex-linked disorders. The same would, presumably, be true of any risks connected with enhancement.

Now this is a startling conclusion, since in paragraph 129 the HFEA state that "[t]he risk of passing on a serious sex-linked genetic condition is a good and, other things being equal, sufficient reason for prospective parents to be offered the options of sex selection." Note that the risks inherent in the procedures of sex selection will still be present where the purpose is to eliminate sex-linked disorders, and so the HFEA's argument seems to be that it is reasonable for parents to expose their future children to risks to their health because the alternative for these parents is to expose different children[12] to greater risks. The HFEA avoid consideration of a third alternative in this context, which is that parents do not need to expose children to significant risks at all. They can use embryo selection or abstain from reproduction. However, since the avoidance of greater risks to different children is not something that can benefit the particular children that will be born as a result of the gender selection, it surely cannot justify exposing these children to risk if the risks are unacceptably high. To talk of greater risks to different children is slightly problematic. If no gender selection is used, either boys or girls may result. Gender selection to eliminate sex-linked disorders usually tries to eliminate males, since they are more likely to be affected. The interesting question is: who benefits? The class of children saved from "risk" by sex selection are at risk of being born with a serious disease. But in some cases such an existence may still be preferable to nonexistence. So they may be saved from the risk of disease at a greater (almost 100%) risk of nonexistence. In such a case it might be argued that only the parents and society benefit from this risk-avoidance strategy and that the welfare of the child to be born has no place in the calculation.

Compare using one child as a bone marrow (or even kidney) donor for a sibling. This would equally be a case in which a child would be required to run risks rather than expose a different child to greater dangers, but I doubt that would be thought obviously consistent with arguments requiring the welfare of the child concerned to be taken into account.[13] I assume that the welfare of "the child who may be born" refers to the welfare of the particular child calculated to be the product of the combination of choices and technology used. The only possible alternative understanding of the meaning of the phrase "welfare of the child to be born" sees it as requiring a eugenic program for reproduction aimed at producing the best of all possible children in the circumstances. This would be an altogether different project, and one, I would suggest, that is even further from the project of the framers of the *HFE Act* 1990 (the act of the U.K. parliament which set up the HFEA) or those who voted for it.

The only way to sustain the idea that seems to be in the heads of the HFEA is to argue that, although the child is exposed to greater risk, such risks are in the interests of the child exposed to them because it is that child's only chance of existence.

It is as if the future children had been offered a bargain: "Here's the deal, you have a chance of coming into existence but only if you accept greater than normal risks—take it or leave it!" A rational embryo or would-be embryo would take the deal, because the alternative is nonexistence. This is the only appeal which makes sense in terms of the interests of "the children who will be born," but I doubt if the HFEA would wish to endorse it because then they would have to do so in the simple gender-selection case.[14]

We should note that this way of thinking of things does not involve the attribution of interests to nonexistent beings (although I see nothing in principle wrong with such an attribution). We may translate the hypothetical deal I have described as if it were put post facto to existing children. We say to them: "OK you exist, but at greater risk than would have been required for other kids to exist. Was it worth it? Was it a good deal?" I imagine they would answer "yes" unless life for them was not worth living.

Now of course there is a sense in which the HFEA are quite right. They want to say that for parents at risk of producing a child with a sex-linked disorder there is an important therapeutic advantage in sex selection. Given that they are going to procreate, it enables parents to have a child with less risk of malformation or disease than available alternatives. For this choice to be ethical we have to judge the risk involved in the gender selection procedure so small as to justify it in terms of dangers to the resulting child, unless we appeal to the argument that asks what a rational embryo would choose—the so-called "nonidentity" argument. This argument, invented by Derek Parfit, shows that reproductive choices which select the child to be born cannot harm that child or do other than promote its welfare unless they create a child with a life not worth living.[15] Remember, in the case of a genetic link with sex-linked disorder the parents get a child without having to risk having a child with a sex-linked disorder, which by hypothesis they do not want. But that is the same in the simple gender-preference case. In that case too, gender selection gives the parents a chance to have a child which is free of a condition (being male or female) that the parents do not want. Of course the parents in one case have a more pressing or serious justification according to some. But this too is a matter of judgment of a considerably problematic nature, for in neither case do the parents have to procreate; they can abstain. The alleged case of "necessity" is predicated upon the procreative imperative. But more argument is needed to show that that imperative involves *mere* procreation as opposed to *chosen* procreation. Which brings us to the HFEA's other arguments concerning choice and to reproductive liberty and the democratic presumption.

Is Gender Selection Harmful to Selected Children?

In his response to my critique of the HFEA's report, Tom Baldwin, Deputy Chair of the HFEA, argues that there is another deeper dimension to "welfare of the child" considerations.[16] Baldwin draws on the work of Jürgen Habermas.[17] Baldwin tells us that

> Habermas argues that an essential ingredient of our conception of ourselves is that we should be able to regard our embodied character (*Leibsein*) as a natural phenomenon, and not something which has been, in some respect, deliberately imposed upon us by others, even by our parents. Of course, we must also recognize that in many ways we have been predisposed by the genes we have inherited from our parents; these predispositions are, however, our bodily inheritance and there is no way in which a human being can be created without some such genetic inheritance. But where a fundamental characteristic such as one's sex has been deliberately selected for, things are different.... There is an inescapable alien intrusion into its subjective sense of itself.

Baldwin offers no reason other than the authority of Habermas as to why "an essential ingredient of our conception of ourselves is that we should be able to regard our embodied character (*Leibsein*) as a natural phenomenon." But again we should be cautious about accepting any of this at face value. Before Darwin, it might, not implausibly, have been asserted that an essential ingredient of our conception of ourselves was that we were created as human beings. Now we know we have evolved, like chimpanzees, in a seamless transition from our common ape ancestor, but most of us seem to have adapted well to this dramatic change in *Leibsein*. The evidence is that human beings are fairly robust and well able to adapt to new conceptions of themselves and their place in the universe. The observations of Galileo and Copernicus were equally, perhaps more, momentous for our conceptions of ourselves and our place in the universe and in the scale of things, but again we seem to have come through. The absorption of Galileo and Copernicus no less than Darwin into our conceptions of ourselves and our place in the world throws doubt upon any conception of a given set of essential ingredients of our conceptions of ourselves.

Although Baldwin accepts the nonidentity of the children that result from different choices, he nonetheless insists in employing it in a way in which it is difficult to make sense of. For example, it "was her parents' choice that their child should be a daughter and in that sense her femininity is indeed imposed on her" seems to imply that their child might not have had her femininity imposed on *her*, that *she* might have been, and might have preferred to be, a boy. How might *she* have escaped this fate and avoided this imposition of femininity?

Reproductive Choice and the Democratic Presumption

In paragraph 132 of the report, the HFEA set out, and commit themselves to, what we have called the democratic or the liberal democratic presumption, a device which, as readers of this book will be well aware, is one of which I heartily approve. They express the presumption thus:

> The main argument against prohibiting sex selection for non-medical reasons is that it concerns that most intimate aspect of family life, the decision to have children. This is an area of private life in which people are generally best left to make their own choices and in which the State should intervene only to prevent the occurrence of serious harms, and only where this intervention is non-intrusive and likely to be effective.

This is a firm and consistent statement of one of the presumptions of liberal democracies, that the freedom of citizens should not be interfered with unless good and sufficient justifications can be produced for so doing. The presumption is that citizens should be free to make their own choices in the light of their own values, whether or not these choices and values are acceptable to the majority. In this report, however, the HFEA simply surrender to the hostility to gender selection of a majority (not of citizens, but of respondents to a consultation which, although it describes itself as "wide," necessarily samples a tiny fraction of the population) and give, in the end, no weight to this important liberal principle and presumption underlying all democratic societies.

The HFEA and Democratic Principles[18]

In the paragraphs following paragraph 132, the HFEA rehearse many of the considerations that were adduced in the public consultation and it is not clear always whether they are endorsing these or merely repeating them. The HFEA rightly dismiss considerations that gender selection may produce a gender imbalance and rely in effect on the following few, and I believe totally inadequate, considerations.

The first is set out in paragraph 139, where they say, "In our view the most persuasive arguments for restricting access to sex selection technologies, beside the potential health risks involved, are related to the welfare of the children and families concerned." The HFEA then gloss this concern for children by noting that, "Children selected for their sex alone may be in some way psychologically damaged by the knowledge that they had been selected in this way as embryos." This is a very tendentious and unwarranted way of putting the point. They produce no evidence, nor indeed could they produce any evidence, that children would be selected for their sex *alone*. This is the point which derives from Kantian ethics, that individuals must be treated as ends in themselves and not as *mere* means. However, it is very difficult to find evidence or even persuasive anecdotes that if people are treated as means they are treated as *mere* means or *exclusively* as means. It is very unlikely that children selected for gender would be selected *solely* for their gender. Indeed it is difficult to understand what that might mean. Also the idea that even if this were the case, they would be so unloved and treated so unacceptably badly that it would cause psychological damage is a piece of reckless speculation for which no evidence is produced and indeed no evidence could be produced. Moreover, the HFEA give no proof or even any reason to suspect that psychological damage would occur. Note, for example, that the process need not be offensive to the child in treating her even as a mere means. Parents will always also care for the child and will want her to feel good about herself; that is (presumably) why they have taken such care over selecting the right gender.

We have already noted the value of experiments in procreation and their connection with freedom of the individual. It is not inconceivable that it will turn out that it is in fact good for children to feel they have been specially selected. Here there is a good case of evidence-based regulation. Allowing freedom for parents and conducting careful follow-up research will show whether or not children are benefited or harmed by such knowledge and the degree of harm or benefit involved. If it turns out to be bad more often than good, parents will presumably stop doing it, and if they don't, *then* is the time to consider legislation.[19]

The HFEA then repeat (or is it "endorse"?) two other considerations, namely "[t]hat such children would be treated prejudicially by their parents and that parents would try to mould them to fulfil their (the parents') expectations. Others saw a potential for existing children in the family to be neglected by their parents at the expense of sex selected children." Well, these so-called dangers may be theoretically possible, but they are hardly realistic. Suffice it to say that for these highly speculative and fanciful dangers (for which no evidence is produced and indeed for which so far as I am aware no evidence exists) to count against the powerful formulation of the democratic imperative would be effectively to deny that imperative any weight or role at all; and indeed this is precisely what the HFEA have done, because in paragraph 147, which is the final statement of the HFEA's justifications for rejecting gender selection and indeed for proposing legislation against it, they say the following:

> In reaching a decision we have been particularly influenced by the considerations set out above relating to the possible effects of sex selection for non-medical reasons on the welfare of children born as a result, and by the quantity of strength of views from the representative sample polled by MORI and the force of opinions expressed by response to our consultation these show that there is very wide-spread hostility to the use of sex selection for non-medical reasons. By itself this finding is not decisive; the fact that a proposed policy is widely held to be unacceptable does not show that it is wrong. *But there would need to be substantial demonstrable benefits of such a policy if the State were to challenge the public consensus on this issue* [my italics].

Thus, the powerful statement of the democratic presumption in paragraph 132 that "the State should intervene only to prevent the occurrence of serious harms" has been converted into the requirement that "there would need to be substantial demonstrable benefits." Here not only has the democratic presumption been turned on its head, but the burden of proof has entirely shifted from the requirement that the State show that its interference is necessary to prevent the occurrence of serious harms to the rather feeble requirement that those who wish to exercise liberty must qualify for this freedom by showing that its exercise provides substantial demonstrable benefits. If this is to be the case, liberty is meaningless and the presumption of liberal democracies is overthrown.

Sex Selection Is a Paradigm of Attempts at Design

Sex selection is, as I have suggested, a good example of the attempt to "design" children where the elements of design are what I have called "morally neutral." Note a number of salient features.

- Gender is genetically determined (so gender selection involves genetic determination).
- Gender is harmless: being a boy or a girl is not bad for you. It is therefore a neutral or personal choice.
- Selection does not, as we have seen, involve shaping the individual in any way, nor can it conceivably make the individual worse off than either (a) she was or (b) she might have been.
- However, it is an example of "design" because it involves choosing between different possibilities or "plans." (Just as when I choose a designer shirt or dress I am indulging my taste in some directions at least, so it is with gender choice.)

The first point needs no further explanation. Let me say something about the claim that gender is harmless. By this I mean simply that it is not harmful to the individual to be a man or a woman. Men and women have existed since humans have and although there have been severe power imbalances between the two genders for most of human history, the damage that this has caused is contingent, not a necessary part of maleness or femaleness. If gender is rightly called a "morally neutral trait," then it cannot be morally wrong to *be* a man or a woman and so it cannot be morally wrong to create a man or a woman. The only remaining question is whether it can be morally wrong to create a man *rather than* a woman or vice versa.[20]

The third element is very important. There is no complaint the "victim" of gender selection can make because for her there was no alternative but never to have existed. "She" could not have been a boy. This is because the boy that might have been selected or created instead of her would not have been "her" only with a different gender. It would not have been a case of sex reallocation. It would simply have involved the creation of an entirely different person.

Although this is hardly ever noticed, the same is true for any significant genetic manipulation that might be made to an embryo or indeed to the gametes prior to conception, if this ever becomes possible. So complaints that parents who would use gender selection are attempting to shape or mold their child are simply incoherent. They may of course be choosing what sorts of children will come into existence, but none of those children have any legitimate or even coherent complaint, for they could not have had an alternative life free of such externally imposed choices.

If the creation or imposition of neutral traits like gender are permissible, then enhancements are surely much less problematic.

10 | The Irredeemable Paradox of the Embryo

We have seen that many of the most dramatic forms of enhancement will use or will be a by-product of therapies and techniques using regenerative medicine and stem cell science to achieve both the therapeutic and the enhancement effects. The ethics of these techniques will continue in part to turn on the legitimacy of sourcing stem cells from embryos and indeed on research using embryos or on embryo-like entities. These are entities either that may be for all intents and purposes embryos, or which those to whom the embryo is "one of us" will regard as embryos or regard as being "possibly" embryos.[1] The moral status of the embryo is therefore of vital importance to the possibility of human enhancement, at least for the foreseeable future. However, many have believed that the question of the moral status of the embryo can be avoided if cells can be derived from entities that are not embryos or, more realistically, that it can be argued are not embryos.

The debate about whether or not these ambiguous entities are embryos is fascinating but, as Katrien Devolder has shown,[2] it is ultimately unsatisfactory as a way of resolving the ethical issues. In this chapter I want to further explore the ambiguous status of the embryo, already hinted at, and to reveal how this ambiguity will ultimately show that the question of the moral status of the embryo is not simply as yet unresolved, but essentially irresolvable in any way that could accord significant moral status to the embryo.

Those who regard the moral status of the embryo as a significant bar to its use as a source of therapeutic, experimental, or enhancement material maintain what may be called the *moral significance thesis*.

The Moral Significance Thesis

The *moral significance thesis* holds that early embryos have equal or similar enough intrinsic worth to that of a person that the intentional or foreseeable destruction of embryos, whether in the sourcing of stem cells or other material from embryos, in therapy, enhancement and medical research, or in any other way is illegitimate.

Against the *moral significance thesis* I shall argue as follows.

- The *moral significance thesis* generates absurd implications connected to embryo splitting (namely, it would be immoral to recombine or split embryos, or to fail to split embryos).
- The determinate defense of the *moral significance thesis*, on which if *X* has the potential to become *Y* then *X* already possesses (much of) *Y*'s value is unconvincing.
- The *moral significance thesis* is not rescued by the "future of value" argument.
- The *moral significance thesis* assumes that early embryos have rights ("intrinsic worth") but they cannot, and most legislators and the law in most jurisdictions and in international courts agrees that they do not.
- The *moral significance thesis* assumes that killing is far, far worse than allowing to die, but that would, among other things, condemn the passengers of flight United 93 to moral obloquy and the intended victims of the terrorists to death (see below).
- Even the supporters of the *moral significance thesis* tend to recognize that embryos lack the kind of worth that would make destroying embryos in embryonic stem cell research wrong. For they procreate (or permit others to do so), although procreation destroys embryos, often for less important purposes. Supporters of the *moral significance thesis* don't even support expanding research into ways to improve

assisted reproductive technology (ART) so that it can eventually replace natural procreation with an embryo-saving method of procreation.

The Ambiguity of the Embryo

The embryo is a deeply, perhaps irretrievably, ambiguous entity, one that defies classification and slips seamlessly between moral, biological, and even mathematical categories. While many features of this ambiguity have been evident for millennia, discussed by Aristotle[3] and in many religious traditions, it is only really with the advent of modern embryology, genome analysis, and stem cell science that the truly radical features of the power of cells to differentiate and specialize have exploded many of the myths and mistakes concerning the embryo and its moral status.[4]

If we start by examining some of the ways in which the ambiguity of the embryo is at its most dramatic, the problems created by this dubious status will become clearer.

Embryo Splitting[5]

When identical twins occur in nature, they result from the splitting of the early embryo in utero and the resulting twins, true clones, have identical genomes. This process can be mimicked in the laboratory, and in vitro embryos can be deliberately split, creating matching siblings, one or both or which can be used for biopsy or research.

This process itself has a number of ethically puzzling if not problematic features. If you have a preimplantation embryo in the early stages of development and split it, let us say into four clumps of cells, each one of these four clumps constitutes a new embryo which is viable and could be implanted with the reasonable expectation of successful development into adulthood (given the dramatic wastage rate of embryos in all human reproduction, see below). Each clump is the clone or identical "twin" of any of the others and comes into being not through conception but because of the division of the early cell mass. Moreover, these four clumps can be recombined into one embryo again. This creates a situation where, without the destruction of a single human cell, one human life, if that is what it is, can be split into four and can be recombined again into one. Did "life" in such a case begin as an individual, become four individuals and then turn into a singleton again? We should note that whatever our answer to this question, all this occurs without the creation of extra matter and without the destruction of a single cell.

Those who think that ensoulment takes place at conception have an interesting problem to account for the splitting of one soul into four, and for the destruction of three souls when the four embryos are recombined into one, and to account for (and resolve the ethics of) the destruction of three individuals, without a single human cell being removed or killed. These possibilities should perhaps give us pause in attributing a beginning of morally important life to a point like conception.[6]

Embryo splitting allows the use of genetic and other screening by embryo biopsy. In embryo biopsy usually one cell is detached from an embryo for genetic testing. If this cell is totipotent (that is if the cell can become any part of the resulting organism including the extraembryonic tissue and membrane or placenta, and this can result in a complete functioning organism), it is effectively an embryo. It is certainly the sort of entity that those who regard the embryo as sacrosanct would believe to be an embryo (see below). Embryo biopsy would then involve the testing of one embryo to ascertain the health and genetic status of the remaining clone or clones (since the rest of the embryo may be further split to multiply cloned siblings). What would the ethical difference be between taking a cell for cell biopsy and destroying it thereafter, and taking a cell to create a clone and then destroying the clone? The answer can only be that destroying the cloned embryo would constitute a waste of human potential, but this same potential is wasted whenever an embryo is not implanted and is also wasted by the cell biopsy used in PIGD, at least if the biopsied cell is totipotent. Even if the cell is pluripotent, there are now techniques (see below) which might restore that cell, thus raising the same moral issues for those who value the embryo.

We should note an odd tension between how we think of the ethics of the destruction of an individual embryo involving cell loss, on the one hand, and destruction of an individual embryo without cell loss, on the other. As we have seen, the process of embryo splitting also allows for recombination. Assuming this technique to be as safe as the creation of clones by cell mass division, if the embryos are recombined following embryo splitting, a number of individual twins have been "destroyed" without the destruction of a single cell. Is such a process more or less ethically problematic?

If, as seems likely, the reason why it is thought objectionable to recombine such clones is the loss of potential human beings, then perhaps it would be considered unethical *not* to split any embryo into as many twins as possible? By so doing we would, after all, maximize just that potential, the loss of which, supposedly, inhibits recombination. If all this has a dizzying effect, it is perhaps because the language that we use misleads us.

We are, as Wittgenstein was so fond of reminding us, often misled by the language we use. To think of these early clumps of cells as "twins" tempts us to think of them as "persons." Imagine an in vitro embryo where all cells are at the totipotent stage; if this bundle of cells were to be split into four clumps of cells, you will have created four (new?) twin embryos. Take three away and destroy them or recombine all four into one and you are in a sense back where you started, having done exactly the same thing in one sense, namely created a single potentially viable embryo with a particular genome. In another sense you have wasted potential experimental material or potentially viable embryos or even killed three human individuals. Yet this waste arguably also occurs whenever a cell mass that could viably be divided is left undivided, or whenever an egg that could be fertilized is left unfertilized. If the recombined embryo, or the surviving quadruplet, is implanted, comes to birth, and grows to maturity, it will have the same genome as it would have had if the division and recombination had never taken place or if its siblings had never been created and disappeared. Will it be the same person as it would have been? Does it have the same identity as it did in its former incarnation? Certainly its life story is different.

It is difficult to analyze the ethics of the possibilities we have just described. In the recombination scenario not a single human cell has been destroyed. In the case of embryo splitting, no new cells or matter have been created and yet three individuals come and go. Certainly no legal rights have been violated and, as we shall see in a moment, no moral rights have either,[7] but have the interests of any individuals been harmed? If these embryos may be said to have an interest in actualizing their potential, then perhaps there may be a sense in which they have been wronged if not harmed.[8]

Rights Revisited

I have claimed that failing to protect embryos does not involve any violation of the rights, of the embryos at least, although the progenitors or others in lawful possession of embryos may have rights at stake. Perhaps a word or two of explanation for this claim is appropriate. There are two main theories of rights: choice theory and interest theory.[9] Choice theory sees rights as securing "the protection and promotion of autonomy or liberty" and interest theory sees rights as serving to further individual well-being or welfare. Clearly on choice theory embryos cannot possess rights because embryos are not autonomous and so their rights cannot be analysed in terms of choices. Nor can you "promote the autonomy of embryos" by trying to ensure that they become autonomous persons because embryos have no autonomy to promote. To be sure the persons they might become will have autonomy, but that is true of the unfertilized eggs and the sperm which will fertilize them and many other entities considered in this chapter which would have to have "rights" for the same reason if the potential for rights conveyed rights. Even according to interest theory embryo rights are problematic, as we have seen, not least because there is no evidence that embryos could experience welfare and therefore they have no welfare interests that can be served. Joseph Raz, for example, suggests that an individual is capable of possessing rights "if and only if . . . his well-being is of ultimate importance." If Raz is right about this then, since embryos have no well-being (for well-being is a state of being experienced as good by the subject of the relevant experiences), they can have no rights. Certainly all early embryos such as those we have been discussing in the preimplantation stage, lacking as they do both a central nervous system and indeed a brain, are incapable of experiences of any sort.[10]

Most legal systems and international human rights laws find no room for attributing rights, particularly the right to life to embryos or fetuses. Two recent cases decided by the European Court of Human Rights have added to a formidable set of precedents confirming that embryos lack both moral and legal personality and therefore have no protectable rights nor interests in life as far as the law of most jurisdictions is concerned.[11]

We must look at potentiality more closely.

Potentiality[12]

We have already noticed how essential the idea of potentiality is in discussing the ethics of using embryos for research and therapy. One feature of human embryos that members of other species do not share is their particular potential, not simply to be born and to be human, but to become the sort of complex, intelligent, self-conscious, multifaceted creatures typical of the human species.

There seem to be two problems with potentiality interpreted as the idea that human embryos or fetuses are morally important beings in virtue of their potential or have a protectable interest in actualizing that potential.

The first objection to protecting individuals because of their potential is logical; acorns are not oak trees, nor eggs omelettes. It does not follow from that fact that something has potential to become something different that we must treat it always as if it had achieved that potential. Unless and until we achieve the possibility of immortality, discussed in chapter 3, all of us share one important and inexorable potentiality—we are all potentially dead meat, but it does not follow that we must be now treated as if we are already dead.

The second difficulty with the potentiality argument involves the scope of the potential for personhood which we considered in chapter 6. If the human zygote (early embryo) has the potential to become an adult human being and is supposedly morally important in virtue of that potential, then what of the potential to become a zygote? Something has the potential to become a zygote, and whatever thing or things have the potential to become the zygote have whatever potential the zygote has. It follows that the unfertilized egg and the sperm also have the potential to become fully functioning adult humans. In addition, it is possible to stimulate eggs, including human eggs, to divide and develop without fertilization (parthenogenesis). As yet it has not been possible to continue the development process artificially beyond early stages of embryogenesis, but if it does become possible safely to create humans parthenogenetically, then the single unfertilized egg, without need of sperm or cloning, would itself have the potential of the zygote. However, from the perspective of those who hold the *moral significance thesis*, this technical problem must be irrelevant, for they believe that anything which is an embryo is protected, regardless of how damaged an embryo it is or how short its life is expected to be.

Cloning by nuclear transfer, which involves deleting the nucleus of an unfertilized egg, inserting the nucleus taken from any adult cell, and electrically stimulating the resulting newly created "embryo" to develop, can, in theory, produce a new human. This was the method used to produce the first cloned animal, Dolly the sheep, in 1997. This means that any cell from a normal human body has the potential to become a new "twin" of that individual. All that is needed is an appropriate environment and appropriate stimulation. The techniques of parthenogenesis and cloning by nuclear substitution mean that neither conception nor fertilization are the necessary precursor processes for the creation of human beings.

To complete this argument we should notice that recent work, brought to philosophical prominence and discussed by Katrien Devolder, explores the consequences of the possibility of returning pluripotent cells to totipotency in the lab and outlines and discusses the ethical and philosophical consequences of being able to do this.[13] For our present purposes Devolder's work highlights a further ambiguity of the embryo, since a pluripotent cell, while it can make any part of the human organism, cannot make the extraembryonic membrane, placenta, etc. It is only totipotent cells which can make absolutely all cell types required not only for the complete organism but also for successful implantation and development. If these possibilities are proved (see below), then there will be elasticity between pluripotent and totipotent cells and it will not be possible to claim that pluripotent cells have permanently lost the capacity to be embryos because they cannot make the required extraembryonic tissue and membrane.

Thus, if the argument from potential is understood to afford protection and moral status to whatever has the potential to grow into a normal adult human being, then potentially every human cell deserves protection. For this important potential is not only possessed by zygotes properly so-called but by whatever has the potential to become an embryo or a zygote or to return to being one.[14]

The account of potentiality given here may be thought to have misrepresented the argument from potential. John Finnis, for example, has argued that "[a]n organic capacity for developing eye-sight is not 'the bare fact that something will become' sighted; it is an existing reality, a thoroughly unitary ensemble of dynamically inter-related primordia of, bases and structures for, development." He concludes that "there is no sense whatever in which the unfertilized ovum and the sperm constitute one organism, a dynamic unity, identity, whole."[15]

However, it is surely the case that *A* has the potential for *Z* if, when a certain number of things do and do not happen to *A* (or to *A* + *N*), then *A* (or *A* + *N*) will become *Z*. For even a "unitary ensemble of dynamically inter-related primordia of bases and structures for development" must have a certain number of things happen to it and a certain number of things that do not happen to it if its potential is to be actualized. If *A* is a zygote, it must implant, be nourished, and have a genetic constitution compatible with survival to term and beyond. Moreover, insistence on a "unitary ensemble," on "one organism," seems also to apply to cloning by nuclear substitution, surely an embarrassing fact. In any adult cell there is a complete single human genome; if treated appropriately, that genome present in the cell nucleus might be cloned. Thus, this method of cloning allows for the "existing reality" of a complete genome which exhibits "dynamic unity, identity, whole[ness]" that the Finnis analysis requires and we can therefore now ascribe potentiality in the Finnis sense to the nucleus of every cell in every body.

The "Future of Value Argument"

The moral importance of drawing attention to the potentiality of something suggests that it is actualizing a particular potential that matters. Our moral concern with what it is that has the potential to become an adult human being would be inexplicable if persons or adult humans did not matter. We are interested in the potentiality argument because we are interested in the potential to become a particular, and particularly valuable, sort of thing. If the zygote is important because it has the potential for personhood, and *that* is what makes it a matter of importance to protect and actualize its potential, then whatever has the potential to become a zygote must also be morally significant *for the same reason*. Those who value potentiality for personhood surely do so not because the potential is contained within "one organism." Rather, they do so because it is the actualization of the potential to become something which has moral importance.

This is one fallacy of Don Marquis's interesting approach to the value of the embryo. For Marquis it is the loss of a "future of value" that he proposes as the wrong done to embryos when decisions result in their failure to survive to experience that future. For Marquis, having a future of value is what makes a creature valuable, gives it a life of moral importance, a life worth saving, a life it is wrongful to end.

But surely that future of value awaits whatever has the potential to become an embryo.[16] Indeed, Marquis's "future of value" argument is really just the potentiality argument reframed, since for Marquis the individual allegedly wronged by the denial of existence is the individual with the potential for a future of value. That is what matters, since without it the embryo has, on the Marquis view, nothing that calls for protection, nothing that makes it valuable.

Although Marquis insists that the future of value must be possessed by an individual, this seems like a stipulation designed to get over the problem of fertilization so that ununited eggs and sperm are not accorded a "future of value" with the exhausting ethic that would entail. But it is difficult to see why (except to save the blushes of Marquis and others) potentiality should be contained within one organism[17] in order to be preserved or actualized or have a future of value. This throws substantial doubt on Marquis's claim that a future of value is possessed only by single individuals. The scientific potential of things other than embryos to become human individuals has recently been demonstrated by developments in stem cell research.[18] Embryonic stem cells (ESCs) are obtained from embryos at the blastocyst stage. A blastocyst consists of two distinct cell types: inner cell mass (ICM) cells, which will become the "embryo proper," the fetus, and later the adult human being; and trophoblast cells, which will form the placental support system necessary for the development of the fetus in the uterus. ESCs are derived from the isolated ICM. So far, it has generally been accepted that human embryonic stem cells (hESCs) have no significant moral status because, just like ICM cells, they are "merely" pluripotent, which means they can form all embryonic but only some extraembryonic tissues. A totipotent cell, i.e., an embryo, can, as we have seen, produce the extraembryonic tissues as well, and can thus result in a whole new individual. This moral division between pluripotent and totipotent cells, however, may not be as sound a criterion as it has seemed to many to be. Scientific evidence suggests that human ICM cells as well as hESCs can also develop into a whole new person. In the mouse it has been proved that the isolated ICM and mouse ESCs derived from it maintain their capacity to form an adult mammal. When they are aggregated with tetraploid embryos—two cell-stage zygotes that have been fused and have twice the normal number of chromosomes—they develop into normal mice.[19] These mice consist only of the ICM cells or the ESCs and not of the tetraploid embryos, which only provide a surrogate trophectoderm. There seems no reason why this same technique would not work with human ICM cells or hESCs. In those for whom the use of tetraploid human embryos as surrogate trophectoderm raises moral issues (involving as it does the instrumental use of embryos), tetraploid embryos could be replaced by trophectodermal cells derived from hESCs.[20] This highlights a further ambiguity of the embryo.[21] If these possibilities are proved, then there will be elasticity between pluripotent and totipotent cells and it will not be possible to claim that pluripotent cells have permanently lost the capacity to form an embryo and a fetus because they cannot make the required extraembryonic tissue and membrane. If hESCs can do everything a totipotent cell or an embryo can, then those who accord full moral status to the embryo should treat hESCs, as well as ICM cells (which have the same potential), as moral equals to the embryo and thus as if they share whatever moral status the embryo has.

Do cells, which could be reprogrammed to be totipotent, have "a future of value" in Marquis's sense? Surely they must because they can be individuated; true, they need a facilitating intervention (but so do all embryos: they need to be implanted in a uterus, nourished, and so on) but we certainly can know which cell has or would have had a future of value.

Consider also a different question: does your individual genome and mine have a future of value? If your genome or mine were to be cloned by cell nuclear transfer to create a zygote, then that genome has a future of value in Marquis's sense.[22] Given that "*n*" things must happen and "*n*" other things must not happen in order for that future of value to be actualized, would not the cloning process count among the community of "*n*" things that must happen? If, as Marquis says, his principle is "victim centered," that is, it is a person-affecting approach to the ethics of embryo sacrifice or manipulation, then in the cloning case it is my genome that provides all of the individuality of the "victim" that Marquis speaks of. In the case of cells that could be reprogrammed to become totipotent, we also and equally know which individual victim has been or will be deprived of a future of value. So, if we ask who or what has been denied a new and different future of value, who is the person affected by the denial of a future of value, the answer is obvious. It is the individual who might have had that future of value, and it is the genome that does most to individuate, to identify, the individual or set of individuals we are talking about. It is this genome that has been denied the opportunity to express itself, and for the individual whose valuable future is at issue this is true in both the reprogramming case and the cloning case.

Interestingly, Robin Lovell-Badge has recently pointed out[23] that

> [t]he ability to derive both sperm and eggs from ES cells entirely in vitro would mean that you could put the two together, obtain blastocysts and then new ES cell lines from these, again all in vitro. In other words, it would be possible to do multigenerational human genetics in the lab, introducing whatever genes or mutations you wanted along the way, without having to worry about men and women liking each other and over a much shorter timescale than occurs naturally. Occasionally you could check what the mature phenotype really looked like after embryo transfer.

Thus, if there is a protectable "interest" in actualizing potential or even a powerful moral reason for doing so, then the consequence is a very demanding ethic and one which would surely require us to actualize all human potential whenever we have an opportunity to do so. This would be a very demanding ethic indeed for women, particularly in the present state of technology. That is not of course a decisive argument against acceptance of the ethic of always attempting to actualize valuable potential. However, those who use this argument as a reason for protecting embryos must in consistency protect whatever has the potential to become an embryo in the same way and to the same extent. This, among many other things, would entail an ethic of maximal procreation, or never knowingly missing an opportunity to create and protect embryos.

It would also entail an ethic of promoting human reproductive cloning on as massive a scale as humanly possible so that the potential locked in the nucleus of almost every cell in every body could be "released" and its potential actualized.

Humans Are Reckless of Embryonic Human Life

A further feature of current attitudes to the embryo which reveal not only the ambiguous status of the embryo but also the ambivalence of most humans toward the precursor forms of themselves concerns the extremely high rate of embryo loss and abnormality in human reproduction. It is doubtful that natural sexual reproduction, with its risk of sexually transmitted disease, its high abnormality rate in the resulting children, and its gross inefficiency in terms of the death and destruction of embryos, would ever have been approved by regulatory bodies if it had been invented as a reproductive technology rather than simply "found" as part of our evolved biology.

Recent research has confirmed abnormality rate for live births associated with sexual reproduction is almost certainly more than 6%[24] but given the moral importance attached to embryos and the fact that embryos are regarded by many as sharing the same moral status with the rest of humankind, it is the tolerated rate of embryo loss that is particularly interesting. Embryo loss in normal sexual reproduction including unprotected intercourse is certainly very high. There is the loss to be associated with every live birth as well as the loss that occurs routinely in unprotected intercourse. To this must be added embryo loss as a direct result of a number of widely used methods of "contraception." In connection with embryo loss associated with sexual reproduction including unprotected intercourse not directly intended to result in conception, Robert Winston gave the figure of five embryos lost for every live birth some years ago in a personal communication.[25] Anecdotal evidence I have received from a number of sources confirms this high figure but the literature is rather more conservative, making more probable a figure of three embryos lost for every live birth.[26] Again, in a personal communication,[27] Henri Leridon confirmed that a figure of three lost embryos for every live birth is a reasonable, conservative figure. Ron Green has suggested that between two-thirds and three-quarters of all embryos do not implant.[28]

Additional embryo loss occurs as a result of the operation of the combined oral contraceptive pill, which has a number of modes of operation, one of which prevents implantation of the embryo at between five and eight days' development. Equally, the so-called "morning after" contraceptive pill also prevents implantation, as does the intrauterine device or "coil."[29] The combined effects of these various contraceptive methods increase the tally of embryo loss as a "side effect" of human sexuality, but it is impossible to arrive at reliable estimates as to the total numbers of embryos involved. Interestingly, most of this embryo loss involves the death of embryos at precisely the stage of development at which stem cells are usually harvested for ESC research, namely between five and eight days' development. This stage is preferred because the cells have not begun to specialize (which takes place at implantation in vivo) and so are still totipotent or pluripotent and therefore can be made to specialize into almost any required cell types.

Those who attempt to have children in the light of these facts and indeed those who have unprotected intercourse, or who use contraceptive methods which risk embryo loss, all must accept that what they are doing or trying to do justifies the creation and destruction of embryos.

In the case of attempts to procreate using sexual reproduction, one obvious and inescapable conclusion is that God and/or nature has ordained that "spare" embryos be produced for almost every pregnancy, and that most of these will have to die in order that a sibling embryo can come to birth. Thus, the willful creation and sacrifice of embryos is an inescapable and inevitable part of the process of procreation. It may not be intentional sacrifice, and it may not attend every pregnancy, but the loss of many embryos is the inevitable consequence of the vast majority of (perhaps all) pregnancies. For everyone who knows the facts of life, it is conscious, knowing, and therefore deliberate sacrifice; and for everyone, regardless of "guilty" knowledge, it is part of the true description of what they do in having or attempting to have children.

The inescapable conclusion is that the production of spare embryos, some of which will be sacrificed, is not unique to ART; it is an inevitable (and presumably acceptable, or at least tolerable?) part of all reproduction.

Both natural procreation and ART involve a process in which embryos additional to those which will actually become children are created only to die. If either of these processes is justified, it is because the objective of producing a live healthy child (and, for some, simply having sex) is judged worth this particular cost. It follows that no one who regards it as acceptable to try to have children (or indeed to have unprotected sex) has any principled objection to the creation and destruction of embryos in a good cause. The only question is how good the cause must be to justify such deliberate embryo destruction. As we have seen, conservatives attracted to Don Marquis's arguments that embryos are to be protected because they possess a future of value (potential) should have no objection to creating embryos that have no chance of survival to experience such a future (but not of course turn them into such embryos). However, those impressed by Marquis's arguments will have to give up procreation because, while sexual reproduction does not usually involve the deliberate or knowing sacrifice of a future of value, Marquis does not seem to think this lack of knowledge or intentionality is entirely exculpatory. Marquis holds that "the future of value theory is a victim-centered theory" and "what is needed for the wrong of killing is an *individual* who is *deprived* of a future of value."[30] But that victim is created and killed by normal sexual reproduction with monotonous regularity. The process of procreation both produces and destroys an individual with a future of value just, for example, as does the creation of embryos for research.

As I have demonstrated elsewhere, Marquis and others would have to favor stem cell research and other forms of embryo experimentation if they afforded the same chance of a future of value or of developing personhood as does sexual reproduction. While Marquis might respond that the concept of "deprivation" involves an alternative and it is not the case that there is a way of saving the embryos lost in sexual reproduction, this may not be or may not remain true.[31] If (or when) assisted reproduction or cloning are, or become, methods of reproduction with a better success rate than sexual reproduction, they will, for all who accept the *moral significance thesis* or the future of value or potentiality arguments, become the only legitimate or ethical methods of reproduction.

As we noted in chapter 9, sexual reproduction is like offering future children the following bargain: "Here's the deal, you have a chance of coming into existence but only if you accept greater than normal risks—take it or leave it!" A rational embryo or would-be embryo would take the deal, because the alternative is nonexistence. Marquis and others would have to accept research on embryos if the research protocols gave research embryos the same chance of a future of value as does sexual reproduction, or they would have to deny the legitimacy of sexual reproduction. As I noted elsewhere:[32]

> [I]t might be claimed [that] embryos produced specifically for research would not rationally choose to participate, for they stand to gain nothing. All research embryos will die and none have a chance of survival. If this argument is persuasive against the production of research embryos it is easily answered by ensuring that the population of research embryos to some appropriate extent have a real chance of survival. One would simply have to produce more embryos than were required for research, randomize allocation to research and ensure that the remainder were implanted with a chance to become persons or have a future of value. To ensure that it was in every embryo's interests to be "a research embryo," all research protocols permitting the production of research embryos would have to produce extra embryos for implantation. To take a figure at random but one that as it happens mirrors natural reproduction and gives a real chance of survival to all embryos, we could ensure that for every, say, 100 embryos produced for research, 10 would be produced for implantation. The 100 embryos would be randomized: 90 for research, 10 for implantation and all would have a chance of survival and an interest in the maintenance of a process which gave them this chance.

Those who accept such destruction as part of a procreative project accept that the creation of new life is a cause good enough to justify such a course of action. Since most people believe that the saving of existing life takes priority over the creation of new life, the use of embryos in the production of lifesaving therapies is clearly justified. Research directed toward lifesaving therapies or the production of those therapies is, it is true, at one further remove from lifesaving, but I believe it must also justify embryo loss if reproduction does.[33]

It is plausible to assume "that those who will the end will the means also."[34] No contemporary therapies are developed without research, and this of course goes as much for lifesaving therapies as for any others. Of course, attempts at lifesaving may always be unsuccessful, as any human attempt may be. If attempts to save lives and the costs of making those attempts are justifiable because of the end to which they are directed, the probability of success is part of the justification. This must also be true of research directed toward lifesaving therapies. The justification is of course weakened in inverse proportion to the probability of success, as it is with more direct forms of lifesaving. The principles are obviously the same. The rest is argument about the likelihood of success.

Those who doubt that attempts to save existing life take precedence over creating new life should consider the ethics of the emergency-room doctor in the following dialogue.[35]

The phone in the doctors' station at the hospital rings.

"Doctor, there is an emergency! You are the only doctor who can help: a life is at risk."

"Nurse, I have more important things to do. My boyfriend and I have just retired to bed intent on procreation, the time is propitious and you will know that creating new life is at least as important as saving existing life (and far more fun!). Tell the patients I am busy with more important things."

Of course, as noted, the probability of success in each case is relevant. We will assume that the chances of the doctor conceiving and subsequently losing the baby prior to birth, taking full account of the potency and fertility of her partner, are balanced by the seriousness of the life-threatening condition of her patient and the doctor's skill levels, etc., so that the probabilities of saving the patient's life and the doctor creating a new life are sufficiently similar for the choice to be between clear alternatives. Those of course who value the embryo from conception equally with all other human beings will be interested only in the probability of successful conception and will discount calculations which include the probability of successful implantation or subsequent miscarriage, etc. For those who regard the embryo as "one of us," female medical and nursing staff of reproductive age will almost always have something more morally important to do than look after patients[36] and at least some male medical and nursing staff (and possibly some patients) may be expected to be among the available procreational partners.[37]

In the case of accepted embryo loss by those not intending to procreate, using perhaps embryo-risking methods of contraception including timing of intercourse, the levels of benefit to be derived from other embryo-costly activities will be proportionately less exacting.

It follows that few sexually active people will be able to object to the creation and sacrifice of embryos in scientific research, human enhancement, or therapy if they consistently apply their principles.

It might be said that there is a difference—those who engage in assisted reproduction create and destroy an unnecessarily high number of embryos. However, those who engage in sexual reproduction are not engaged in the destruction of embryos at a greater rate than is required for the outcome they seek. It would be interesting to know whether, if creating a single embryo by IVF became a reliable technique for procreation, those from the rather inappropriately termed "pro-life" position would feel obliged to use this method rather than sexual reproduction because of its embryo-sparing advantages. (Inappropriately termed "pro-life" because those who regard themselves as "pro-life" so often support positions which can only be thought of as anti-life and which moreover are profligate of human life and safety.[38])

Flight "United 93"[39]

On September 11, 2001, passengers on flight "United 93" are reliably believed to have overcome hijackers and forced a hijacked plane to crash into a field in Pennsylvania, so forestalling the attempt to target a highly populated and high-profile building, but killing everyone on board. Such an act, while defending the victims in the "target of choice" did involve killing the innocent passengers and crew. Not all the passengers could conceivably have consented to the takeover of the plane and the deliberate bringing about of the earlier crash landing, so there must inevitably have been the decision by some passengers deliberately to sacrifice nonconsenting others. True, they would all almost certainly have died anyway, but they were killed before they otherwise would have been. Although their deaths were probably inevitable, the deliberate (if voluntary for some) killing must have offended against the sanctity of life doctrine, for, on a usual interpretation of pro-life positions, killing the innocent who are posing no threat is not legitimate, however noble the justification. And killing earlier than an inevitable subsequent death is still killing, otherwise euthanasia would be less problematic than it appears to some. So this widely praised act, which probably saved many innocent lives, must, for pro-lifers, be one of pure wickedness on the part of the passengers who resisted the hijack if their intent was to crash the plane before it could reach its target and not exclusively to overpower the hijackers.[40] The espousal by successive popes of a rigorous hostility to condom use in the face of the continuing AIDS pandemic is estimated to have cost millions of lives and untold misery.[41]

It looks as though there would indeed be a strong moral obligation to abandon natural procreation and use only embryo-sparing ART. Indeed, if such an improvement in IVF occurred, this would seem to make using IVF mandatory for those who believe the embryo is one of us. And it is interesting that so-called "pro-lifers" are not (apparently) investing heavily in IVF to this end in the hope that sexual reproduction could eventually be entirely replaced by an embryo-sparing method of reproduction.

What follows from all this? It is difficult to see how most people could live lives that are today accepted as normal while maintaining a strict "pro-life" position or by acting consistently to protect embryos. The alternatives seem clear. We, humankind, must surely accept that human embryos are deeply ambiguous and problematic entities of a kind whose lives or "dignity" simply cannot be protected in ways consistent either with other values that we hold or indeed with the continued existence of the human species. The alternative is consistent but bleak; it involves the strict avoidance of all acts which would violate the sanctity of life of embryos. This would of course include almost all human procreation and certainly all sexual reproduction.

Personal Identity

Making People Better without Changing Their Identity or Interfering with It

Any radical changes in individuals that significantly affect their self-image or the way in which they can lead their lives are likely to have impacts on identity. This may be true whether the changes take place at the embryonic stage or before, or indeed at any time during life. How serious the impact has to be before we question whether as a result the individual concerned has ceased to be the person that she was is a matter of considerable controversy. Derek Parfit has famously claimed that each person is less connected than most of us believe:

> Strong connectedness is *not* a transitive relation. I am now strongly connected to myself yesterday, when I was strongly connected to myself two days ago, when I was strongly connected to myself three days ago, and so on. It does not follow that I am now strongly connected to myself twenty years ago. And this is not true. Between me now and myself twenty years ago there are many fewer than the number of direct psychological connections that hold over any day in the lives of nearly all adults. For example, while most adults have many memories of experiences that they had in the previous day, I have few memories of experiences that I had on any day twenty years ago....
>
> Since strong connectedness is not transitive, it cannot be the criterion of identity. And I have just described a case in which this is clear. I am the same person as myself twenty years ago, though I am not now strongly connected to myself then.[42]

Curing illnesses of and preventing diseases for people who already exist do not usually raise problems of identity. Enhancements of people's quality of life will take place, hopefully as a result of enhancement technologies, but people do not normally change drastically enough to raise questions as to whether or not they have remained the same person when they recover from a cold or even from cancer. A formerly totally paralyzed person could be an exception. Is someone who can walk and talk for the first time a different person? Some mental ailments could fall under this category as well: "it is still me, only I do not hear voices in my head any more"[43] or "now that I no longer hear voices I am a completely different person." Similarly, with enhancements, if we can change people so that their intelligence is of a different order of magnitude than any existing intelligence or we can improve memory or physical powers in ways that constitute step changes, issues of identity may arise. But again it is unlikely that the identity issues will be crucial or even relevant in determining the ethics of the proposed enhancements because, as we saw in chapter 4, failure of identity does not necessarily vitiate the rational motive for enhancement.

Making People Better Probably without Changing Their Identity but Possibly Interfering with It

Prenatal (gene) therapy might eventually enable us to remove a seriously disabling condition before implantation of the embryo. Here, such an enhanced individual might well feel disposed to say, on hearing what life would have been like but for the intervention, "I would have been a different person without the intervention."

Consider serious mental illness or impairment, and cognitive and mood enhancements by surgery, therapy, or medication or even gender reassignment. All of these possibilities might lead the recipient to say: "The real me has now surfaced thanks to Prozac" or "I feel 'myself' for the first time in my life!"

How we would or ought to feel about the personal identity issues in any of these cases is difficult to know in the abstract and I strongly suspect that we would need to consider them on a case-by-case basis. However, while the personal identity issues are fun philosophically, they don't seem, to me at least, to have any ethical impact at all.

Indeed, such puzzles seem to be a gross form of philosophical indulgence at the expense of moral decency. Consider, as we noted in chapter 4, the following suggestion by Leon Kass:

> For to argue that human life would be better without death is, I submit, to argue that human life would be better being something other than human.... The new immortals, in the decisive sense, would not be like us at all. If this is true, a human choice for bodily immortality would suffer from the deep confusion of choosing to have some great good only on condition of turning into someone else.[44]

Kass's argument seems to suggest, as we have noted, that since the (current) essence of being human is to be mortal, immortals would necessarily be a different type of being and therefore have a different identity. This is the time to note again that while there is a sense in which this is true, there no sense in which it would be irrational to want to change identity to the specified extent. Moreover, it is clearly ethically problematic in the extreme to suggest that the highly intellectualized "identity" issues offer any moral reasons for the individual to forgo benefits on the absurdly scholastic ground that, at least for philosophers and the philosophically minded, this might raise issues of identity that are unlikely to be susceptible to definitive resolution. As noted above, someone who had been profoundly disabled from birth (blind say, or crippled) and for whom a cure became available in his or her mid-forties would become in a sense a different person. They would lead a different type of life in many decisive ways. It does not follow that the blind or crippled individual has no rational motive to be cured because the cure would involve, or would risk involving, the sacrifice of personal identity. It would be both odd and cruel to say to them, as Kass presumably would have us do, "it is deeply confused to want to cease to be disabled, because then you will no longer exist."

Related paradoxes are also raised for the possibility of time travel. It is, for example, standardly claimed that time travel is not only physically but somehow logically impossible because the idea entails a situation that involves an impossibility. Specifically, if we could travel into the past we could encounter our own genetic parents (or grandparents, etc.) before our conception and prevent that event (usually, it is suggested, by killing one or both of our parents). People who raise this objection seem to think that time travelers are all incipient matricides or else extremely careless! What would actually happen in such an event is of course unknowable, like encounters between irresistible forces and immovable objects or indeed like the question beloved of atheists the world over: if God is omnipotent, can he make a stone that is so heavy he cannot himself lift it? Douglas Adams is among many fiction writers to have considered this and related "impossibilities":

> One of the major problems encountered in time travel is not that of accidentally becoming your own father or mother. There is no problem involved in becoming your own father or mother that a broadminded and well adjusted family can't cope with. There is also no problem about changing the course of history—the course of history does not change because it all fits together like a jigsaw. All the important changes have happened before the things they were supposed to change and it all sorts itself out in the end. The major problem is quite simply one of grammar.[45]

Adams's response does not of course dissolve the paradox, but it reminds us that there are so many unknowns about such scenarios that not much can be learnt from considering them as possibilities. We do not, in short, know what the imagined scenarios will really involve or how the concepts that seem problematic would be analyzed if the imagined events occurred, or what would lead us to say that time travel had been achieved or not. The problem may, as Douglas Adams suggests, be one of grammar. Michael Crichton has also written about time travel.[46] In his scenario, such travel involves visiting one or more of an infinite number of parallel universes. Whether killing your parents in a parallel universe does or does not involve your own nonexistence and hence (possibly) the dissolution of the enterprise in a puff of logic is a question for which the resolution must wait upon further empirical research.

The worries about enhancement which focus on issues of identity or psychological continuity, while fascinating, do not, it seems to me, materially affect the ethics of the social policy dimensions of enhancement. For my own part, as Jonathan Glover[47] once memorably said, "I would be glad of the chance to sample a few million years and see how it went." The identity problems that this may give me and others I will happily leave to time and psychiatry to sort out. If I eventually turn out to have personality disorders beyond the wildest dreams of psychiatry this will be good for psychiatrists and may or may not be bad for me. However, this does not seem to be a powerful reason for me to deny myself or for you, paternalistically, to deny me the opportunities that may, but far more likely may not, have these dire consequences.

11 | The Obligation to Pursue and Participate in Research[1]

In his *Life of Galileo*, Bertolt Brecht gives a memorable insight into the justification of science to Galileo. Talking to Andrea Sarti, his former student and colleague, who is about to smuggle the *Discorsi*, Galileo's heretical treatise on mechanics and local motion out of Italy, Galileo says[2]:

> The battle for a measurable heaven has been won thanks to doubt; but thanks to credulity the Rome housewife's battle for milk will be lost time and time again. Science, Sarti, is involved in both these battles. A human race which shambles around in a pearly haze of superstition and old saws, too ignorant to develop its own powers, will never be able to develop those powers of nature which you people are revealing to it. To what end are you working? Presumably for the principle that science's sole aim must be to lighten the burden of human existence. If the scientists, brought to heel by self-interested rulers, limit themselves to piling up knowledge for knowledge's sake, then science can be crippled and your new machines will lead to nothing but new impositions. You may in due course discover all that there is to discover, and your progress will nonetheless be nothing but a progress away from mankind. The gap between you and it may one day become so wide that your cry of

> triumph at some new achievement will be echoed by a universal cry of horror. . . .
>
> Had I stood firm the scientists could have developed something like the doctors' Hippocratic oath, a vow to use their knowledge exclusively for mankind's benefit. As things are, the best that can be hoped for is a race of inventive dwarfs, who can be hired for any purpose.

The possibility of human enhancement highlights not only an important justification for pursuing the science that will yield human enhancements, but one of the strongest justifications for the entire enterprise of science, one moreover that many scientists who take the "knowledge for its own sake" approach to science have firmly rejected. At the end of this investigation into the ethics of human enhancement it is not inappropriate to consider the ethical imperatives behind the science that makes, or perhaps we should more cautiously say "may continue to make," human enhancement a reality.

We began with the idea that human enhancement is good by definition, just as a benefit must obviously be beneficial. This is trivially true, but enhancements are also good of course because those things we call enhancements do good: they make us better, not simply by curing or ameliorating our ills, but because they make us better people, less the slaves to illness and premature death, less fearful because we have less to fear, less dependent, not least upon medical science and on doctors. For these and other reasons we have suggested that enhancements are a moral duty.

We have seen that many of those who oppose enhancement wish us to accept the world as it is, to accept our limitations, even to take pride in them and rejoice at them. In Michael Sandel's words "appreciating the gifted quality of life constrains the Promethean project and conduces to a certain humility."[3] Sandel and other objectors to enhancement want us to be humble . . . ever so humble. They also want us to foster "an openness to the unbidden" and an acceptance of the limitations that come, always unbidden, into our lives. Again and of course, when something is unavoidable or inevitable there can be wisdom in acceptance, in making the best of a bad lot. But science and technology are all the time changing our conception of what is inevitable and what is possible and they can do this only because science and technology are ongoing activities. They have a place in our world and in our societies, not simply because people are naturally intellectually curious, but because science and technology do palpable good. This good will not occur without the practice of science but this cannot rely on the voluntary activities of the curious alone. It requires support, endorsement, and protection. In this chapter I consider the moral case for science research, for the intellectual activity and practical research which generates not only most of the enhancements we have been considering but indeed most of the non-enhancing therapies and practices which continue to make the world a safer and a better place.

At the heart of the difference between my own approach to enhancement and that of those who oppose it is a view of both the nature of philosophy and the nature of science.

In terms of philosophy, I believe, as Karl Marx remarked (but it might as easily have been Plato), "The philosophers have only interpreted the world, in various ways; the point, however, is to change it,"[4] and I share with Bertolt Brecht the idea that the purpose of science is necessarily also an interested not simply a disinterested purpose—it is a moral purpose.[5] Of course we can be simply curious if we like, but the justification for science and for philosophy is not simply in terms of the disinterested pursuit of knowledge but in terms of the benefits that such a pursuit can bring. There is a universal responsibility to pursue the good, for ourselves and others. With it comes a responsibility to be analytic and autonomous, to decide what is good and to seek it out.

It is perhaps obvious (although this book has provided arguments that show the truth of these obvious ideas)[6] that, granted enhancements do good, are beneficial, serve our interests, and protect us from harm, we always have moral reasons to pursue them. But what of the so-called "blue skies" research, that is, research which is curiosity driven and which does not appear to have an immediate practical or beneficial outcome that can lead to techniques of enhancement becoming a reality? Is there a moral obligation to undertake, support, and even to participate in serious scientific research? If there is, does that obligation require not only that research be undertaken but also that "we" as individuals and "we" as societies be willing to support and even participate in research where necessary? Is there, in short, a strong moral argument not only to avail ourselves of such enhancements as are to hand but also to carry out, fund, and even participate in the research required to generate new enhancing technologies and practices?

Thus, by far the overwhelming answer given to this question has been "no," and research has almost universally been treated with suspicion and even hostility by the vast majority of all those concerned with the ethics and regulation of research. The so-called "precautionary approach"[7] sums up this attitude, requiring dangers to be considered more likely and more serious than benefits and assuming that no sane person would or should participate in research unless they had a pressing personal reason for so doing or unless they were motivated by a totally impersonal altruism. International agreements and protocols (The Declaration of Helsinki[8] and the CIOMS Guidelines[9], for example) have been directed principally to protect individuals from the dangers of participation in research and to ensure that, where they participate, their full informed consent is assured. The overwhelming presumption has been and remains that participation in research is a superogatory and perhaps even a reckless act, not an obligation.

Suspicion of doctors and of medical research is well-founded. In the modern era it was revived by the aftermath of the Nazi atrocities and by the original Helsinki declaration prompted, although rather belatedly, by the Nazi doctors' trial at Nuremberg.[10] More recently, it has been fueled by further examples of extreme medical arrogance and paternalism, for example, the Tuskegee Study of Untreated Syphilis,[11] in which 412 poor African-American men were deliberately left untreated from 1932 to 1972 so that the natural history of syphilis could be determined.[12] Even when penicillin became known to be effective against syphilis, they were left untreated. More recently, in the United Kingdom, a major scandal caught the public imagination and reflected serious medical malpractice; it involved the unauthorized and deceitful post mortem removal and retention of organs and tissue from children.[13]

These and many other cases seem to provide ample justification for the presumption of suspicion of and even hostility to medical research, and scientific research more generally. However, vigilance against wrongdoing is one thing, the inability to identify wrongdoing with the result that the good is frustrated and harm caused is quite another.

When we ask whether or not there is a moral obligation to support and even to participate in serious scientific research, we need first to be clear that we are talking of research directed toward preventing serious harm or providing significant benefits to humankind. In all cases the degree of harm or benefit must justify the degree of burden on research subjects, individuals, or society. This balance will be explored below. Of course, the research must also be serious in the sense that the project is well-designed and with reasonable prospect of leading to important knowledge that will benefit persons in the future.[14]

The justification of science must be seen in terms of the good that it does. The problem of course is that we do not always know and cannot sometimes even anticipate the applications of scientific enquiry. For this reason, as we have noted, it is necessary to defend and indeed actively to promote even so-called "blue skies" research which is serious, well-designed, and which investigates fundamental problems. Of course the risks we may be prepared to run and the costs which it would be rational to bear when we have no idea about possible outcomes will be proportionally more modest. In an important report submitted to the National Bioethics Advisory Commission (NBAC) of the United States, Dr. David Korn[15] meticulously documented the ways in which archive samples, retained organs, and tissue, often kept simply in case it should prove useful for research, have contributed massively and often unexpectedly to medical advances. This report contains very many examples of research where the particular beneficial use could not have been, and was not in fact, anticipated in advance.

Two separate but complementary lines of argument underpin a powerful obligation to pursue, support, and participate in scientific research.

Do No Harm

The first is one of the most powerful obligations that we have: the obligation not to harm others. Where our actions will or may probably prevent serious harm, then if we can (reasonably, given the balance of risk and burden to ourselves and benefit to others), we clearly should act because to fail to do so is to accept responsibility for the harm that then occurs.[16] This is the strong side of a somewhat weaker, but still powerful, duty of beneficence: our basic moral obligation to help other people in need. This is sometimes called "the rule of rescue."[17] We tend to think of rescues as dramatic events with heroes snatching victims from the jaws of death. However, rescue occurs whenever people faced with threats to their lives or health receive assistance which mitigates these needs or their effects. Most, if not all, diseases create needs, in those who are affected and in their relatives, friends, and carers, and indeed in society. Because medical research is a necessary component of relieving that need in many circumstances, furthering medical research becomes a moral obligation. This obligation involves supporting research in many ways, for instance, economically, at the personal, corporate and societal levels, and indeed politically it also involves physical participation in research projects.

We have seen that there is a continuum, an overlap or potent analogy between treating or curing dysfunction and enhancing function. This continuum has two dimensions. In one sense, the withholding of a benefit that could be conferred harms the potential recipient. It is always in our interests to receive a benefit, against our interests not to receive a benefit. If the potential recipient knows about the possibility of receiving the benefit, she will probably wish to receive it, so that not only her interests but also her desires, wants, and choices are engaged by the possibility of receiving benefits. Often also, as we have noted, the very same actions or procedures that constitute therapies for some will offer enhancement to others. These two dimensions of the continuum between ameliorating dysfunction and enhancing function mean that a decision to withhold benefit is always damaging: always something a decent person has a moral reason not to do. Whether those moral reasons are decisive will, of course, depend on many other things, including an assessment of the risks and costs of conferring the benefit.

Fairness

Second, the obligation also flows from an appeal to basic fairness. This is sometimes expressed as an appeal to the unfairness of being a "free rider." We all benefit from the existence of the social practice of medical and scientific research. Many of us would not be here if infant mortality had not been brought under control, or antibiotics had not been discovered and made usable. Most of us will continue to benefit from these and other medical advances (and indeed other advances such as clean drinking water and sanitation). Since we accept these benefits, we have an obligation in justice to contribute to the social practice which produces them. We may argue that since we could not opt out of advances that were made prior to our becoming capable of autonomous decision making we are not obliged to contribute. But it may still be unfair to accept the benefits of science and such acceptance implies also that we will forgo the fruits of any future advances.[18] For example, we bear duties of reciprocity to our parents for their care of us although we did not choose our parents nor the care they gave us in early life. Few, however, are willing to reject benefits to which we have not contributed, and even fewer really willing to forgo benefits that have been created through the sacrifices of others when their own hour of need arises! Reciprocity, however, is not confined to circumstances in which a direct return can be made to those who have previously contributed. Reciprocity is not "repayment," it is more akin to mutuality, solidarity, or fellowship. The idea is one of a community which accepts mutual responsibility, not least because such mutuality has been accepted by others and because all have benefited from the actions or forbearance of others.

It should be clear how what I am claiming relates to the principle which is sometimes called the "principle of fairness," developed by Herbert Hart and later by John Rawls.[19] That principle may be interpreted as saying "those who have submitted to ... restrictions have a right to similar acquiescence on the part of those who have benefited from their submission."[20] Here I am not suggesting an *enforceable* obligation to participate based on fairness, although such an enforceable obligation would, as we shall see, certainly in some circumstances be justified by the argument of this chapter. Nor am I proposing any *right*, possessed by those who participate, to similar acquiescence on the part of those who benefit. However, being a free rider *is* unfair and people always have a moral reason not to act unfairly. This moral reason is probably enough to justify an enforceable obligation but we do not have to use compulsion as a strategy of first resort. It is surely powerful enough, however, to rebut some of the presumptions against an obligation to support and participate in research.

There may be specific facts about me and my circumstances that absolve me from the obligation to be a research subject in a given situation. This could be the case if I have just participated in other burdensome experiments and there are other potential research subjects who have not done so, or if participation would create excessive burdens for me that it would not create for other potential participants. This does not show that the general obligation we have identified does not exist, just that it, like most other or perhaps all moral obligations, can be overridden by other moral considerations in specific circumstances.[21]

The Moral Imperative for Research

We all benefit from living in a society, and, indeed, in a world in which serious scientific research is carried out and which utilizes the benefits of past research. It is both of benefit to patients and research subjects and in their interests to be in a society which pursues and actively accepts the benefits of research and where research and its fruits are given a high priority. We all also benefit from the knowledge that research is ongoing into diseases or conditions from which we do not currently suffer but to which we may succumb. It makes us feel more secure and gives us hope for the future, for ourselves and our descendants, and others for whom we care. Finally, and most importantly given our present concerns, we have all benefited from the many enhancing practices, methodologies, technologies, and evolutionary strategies that have enabled us to adapt to a changing environment and develop powers and capacities scarcely imaginable to our ancestors, human or ape.

If this is right, then you and I have a strong general interest that there be research (and that it be well-founded), not excluding, but not exclusively, research on ourselves and on our condition or indeed on conditions which are likely to affect me and mine, you and yours. All such research is also of clear benefit to you and me. A narrow interpretation of the requirement that research be of benefit to the subject of the research is therefore perverse.[22]

Moreover, almost everyone now living, certainly everyone born in high-income, industrialized societies, has benefited from the fruits of past research. We all benefit, for example, either from having been vaccinated against diseases like polio, smallpox, and others or from the so-called "herd" immunity because others have been vaccinated; or we benefit (as in the case of smallpox) from the fact that the disease has actually been eradicated. To take another obvious example, almost at random, we all benefit from the knowledge of connections between diet, exercise, and heart disease. This knowledge enables us to adopt preventive strategies and indeed enhancement strategies and gives us ways of calculating our level of personal risk.

In view of these considerations there is a clear moral obligation to participate in medical research or any research that has reasonable prospect of enhancing our lives or our material condition in certain specific circumstances. This moral obligation is, as we have seen, straightforwardly derivable from either of two of the most basic moral obligations we have as persons. This obligation is importantly not confined to purely therapeutic research but also involves all beneficial research including research into human enhancement. Of course no one should be obliged to take disproportional risks, and participation in research where there is serious (that is, a high) probability of risk to life or health should not be mandatory.[23]

How Should We Understand What Is "In Our Interests"?

This entails that there are circumstances where an adult, competent person ought to participate in research, even if participating is not in his or her best interests narrowly defined. If I am asked to give a blood sample for a worthwhile research project, or if I am asked if tissue removed during an operation may be retained for research or therapeutic use, I may have to think in the following way: in the case of giving the blood sample I may say to myself "I hate needles and the sight of my own blood!" Equally, with retained tissue or organs I may feel that since I understand little of the future uses for my tissue it would be safer to say "no."

In each case we will suppose that the disease being investigated is not one that I or anyone I know is likely ever to get, so giving this blood sample or allowing the use of excised tissue is not in my best interests narrowly conceived. In this situation doing what is best, all things considered, therefore seems to entail not doing what is best for myself, not pursuing my own best interests. But this is not really so. Some of my main interests have not been identified and taken into account in this hypothetical train of thought. One of these is my interest in taking myself seriously as a reflective moral agent, and my interest in being taken seriously by others. Identifying my moral obligations, and acting on them, is not contrary to my interests, but is an integral part of what makes me a moral agent.[24] But more importantly, as we have seen, I do have a powerful interest in living in a society and indeed in a world in which scientific research is vigorously pursued and is given a high priority.

Research into enhancing procedures is almost always, as we have seen, also therapeutic because almost all conceivable enhancement procedures will also have nonenhancing or not necessarily enhancing uses, such as where procedures that will enhance normal brains might repair damaged brains and vice versa. But, in cases where research was aimed purely at enhancement, the moral imperative would be equally clear, since we all may expect to benefit personally from enhancements, though not all of us may ever be victims of any particular disease or injury.

Do Universal Moral Principles Deny This Claim?

A number of the most influential international protocols on science research seem to contradict the claims made so far and we must now examine these more closely.[25] One of the most widely cited principles is contained in a crucial paragraph of The World Medical Association Declaration of Helsinki:[26]

> **Paragraph 5**. In medical research on human subjects, considerations related to the well-being of the human subject should take precedence over the interests of science and society.

This paragraph is widely cited in support of restrictions on scientific research and is interpreted as requiring that all human subject research is in the narrowly conceived interests of the research subjects themselves; this article of faith has become almost unchallengeable.

We need first to examine more closely the idea of what is or is not in someone's interests.[27] We should note at the outset that what is or is not in a particular individual's interests is an objective matter. While subjects have a special role to play in determining this, we know that human beings are apt to act against their own interests. Indeed, the idea of respect for persons which underpins this guideline has two clear and sometimes incompatible elements, namely, concern for welfare and respect for autonomy. Because people often have self-harming preferences (smoking, drug abuse, selfless altruism, etc.) they are sometimes bad judges of their own interests.

The interests of the subject *cannot* be paramount nor can they automatically take precedence over other interests of comparable moral significance. The claim that the interests of the subject cannot be paramount involves a straightforward mistake: being or becoming a research subject is not the sort of thing that could conceivably augment either someone's moral claims or, for that matter, her rights. *All* people are morally important and, with respect to one another, each has a claim to equal consideration. No one has a claim to overriding consideration. To say that the interests of the subject must take precedence over those of others, if it means anything, must be understood as a way of reasserting that a researcher's narrowly conceived professional interests must not have primacy over the human rights of research subjects.[28] However, as a general remark about the obligations of the research community, the health care system, or of society or indeed of the world community, it is not sustainable.

This is not of course to say that human rights are vulnerable to the interests of society whenever these can be demonstrated to be greater. On the contrary, it is to say that the rights and interests of research subjects are just the rights and interests of persons and must be balanced against comparable rights and interests of other persons. In the case of medical research, the contrast is not between vulnerable individuals on the one hand and an abstract entity like "society" on the other, but rather between two different groups of vulnerable individuals. The rights and interests of research subjects are surely not served by privileging them at the expense of the rights and interests of those who will benefit from research. Both these groups are potentially vulnerable, neither is obviously prima facie more vulnerable or deserving of special protection.

Indeed, often there is no mutual exclusivity between the two groups, research subjects often also benefiting from research and often being also the most immediate beneficiaries.

It is important to emphasize that the point here is not that there is some general incoherence in the idea of sometimes privileging the rights and interests of particularly vulnerable groups in order to guarantee to them the equal protection that they need and to which they are entitled. Rather, I am suggesting two things. The first is that all people have equal rights and entitlement to equal consideration of interests. The second is that any derogation from a principle as fundamental as that of equality must be justified by especially powerful considerations.

Finally, although what is or is not in someone's interests is an objective matter about which the subject herself (or himself) may be mistaken, it is usually the best policy to let people define and determine "their own interests." While it is of course possible that people will misunderstand their own interests and even act against them, it is surely more likely that people will understand their own interests best. It is also more respectful of research subjects for us to assume that this is the case unless there are powerful reasons for not doing so.[29]

Is There an Enforceable Obligation to Participate in Research?

It is widely recognized that there is clearly sometimes an obligation to make sacrifices for the community or an entitlement of the community to go so far as to deny autonomy and even violate bodily integrity in the public interest and this obligation is recognized in a number of ways.[30]

There are a perhaps surprisingly large number of cases where we accept substantial degrees of compulsion or coercion in the interests of those coerced and in the public interest. For example: in limiting access to dangerous or addictive drugs or substances; control of road traffic including compulsory wearing of car seat belts; vaccination as a requirement, for example, for school attendance or travel; screening or diagnostic tests for pregnant mothers or for newborns; genetic profiling for those suspected of crimes; quarantine for some serious communicable diseases; compulsory military service; detention under mental health acts; safety guidelines for certain professional activities of HIV positive people; and compulsory attendance for jury service at criminal trials. Universal education for children, requiring as it does compulsory attendance in school, is another obvious example. All these involve some denial of autonomy, some imposition of public standards even where compliance is not based on the competent consent of individuals. However, these are clearly exceptional cases in which overriding moral considerations take precedence over autonomy. Might medical research be another such case?

Mandatory Contribution to Public Goods[31]

The examples cited above demonstrate a wide range of what we might term "mandatory contribution to public goods." I will take one of these as a model for how we might think about participation in science research.[32]

All British citizens between the ages of eighteen and seventy[33] are liable for jury service. They may be called and, unless excused by the court, must serve. This may involve a minimum of ten days but sometimes months of daily confinement in a jury box or room, whether they consent or not. However, although all are liable for service, only some are actually called. If someone is called and fails to appear, they may be fined. Most people will never be called but some must be if the system of justice is not to break down. Participation in, or facilitation of, this public good is mandatory. There are many senses in which participation in vaccine or drug trials involve features relevantly analogous to jury service. Both involve inconvenience and the giving up of certain amounts of time. Both are important public goods. It is this latter feature that is particularly important. Although jury service (or compulsory attendance as a witness) is an integral part of "due process," helping to safeguard the liberty and rights of citizens, the same is also true of science research. Disease and infirmity have profound effects on liberty and, while putting life-threatening criminals out of circulation or protecting the innocent from wrongful imprisonment is a minor (numerically speaking) product of due process, lifesaving is a major product of science research. If compulsion is justifiable in the case of due process, the same or indeed more powerful arguments would surely justify it in the case of science research.

Of course "compulsion" covers a wide range of possible measures. Compulsion may simply mean that something is legally required, without there being any penalties for noncompliance. Such legal requirement may be supported by various penalties or incentives, from public disapproval and criticism, fines, or loss of tax breaks on the one hand, to imprisonment or forcible attendance or participation further along the spectrum on the other. To say that it would be legitimate to make science research compulsory is not to say that any particular methods of compulsion are necessarily justified or justifiable. While it seems clear that mandatory participation in important public goods is not only justifiable but also widely accepted as justifiable in most societies, as the examples above demonstrate, my own view is that voluntary means are usually best and that any form of compulsion should normally be a last resort to be used only when consensual means fail or where the need for a particular research activity is urgent and of overwhelming importance. If the arguments of this chapter are persuasive, compulsion should not be necessary and we may expect a climate more receptive to both the needs and the benefits of science. However, to point out that compulsion may be justifiable in some circumstances in the case of science research establishes that a fortiori less stringent means are justifiable in those circumstances.

I hope it is clear that I am not here advocating mandatory participation in research, merely arguing that it is in principle justifiable, and may in certain circumstances become justified in fact. There is a difference between ethics and public policy. To say that something is ethical and therefore justifiable is not the same as saying it is justified in any particular set of circumstances, nor is it to recommend it nor yet to propose it as a policy for immediate or even for eventual implementation. For example, if I say that prostitution is justifiable and should be permitted, I am not necessarily encouraging its use either for sex workers or "punters," nor yet as a career for my niece or nephew. Consensual participation is usually preferable and persuasion by a combination of evidence and rational argument is almost always the most appropriate way of achieving social and moral goals. This chapter is an attempt to do precisely this. I believe, for example, that conscription into the armed forces is justifiable, but I am not recommending, still less advocating, its reintroduction into the United Kingdom at this time. The distinction between ethical argument and policy proposal is crucial but is almost always ignored, particularly by the press and news media that report on these matters. Here I am intending to do ethics; this chapter is not a policy proposal, although it does contain one policy proposal, which we will come to in due course.

If I am right in thinking that research is a public good, that may in extremis justify compulsory participation, then a number of things may be said to follow.

- It should not simply be assumed that people would not wish to act in the public interest, at least where the costs and risks involved are minimal. In the absence of specific evidence to the contrary, if any assumptions are made, they should be that people are public spirited and would wish to participate.[34]
- It may be reasonable to presume that people would not consent (unless misinformed or coerced) to do things contrary to their own interest and to that of the public. The reverse is true when (as with vaccine trials) participation is in both personal and public interest.
- If it is right to claim that there is a general obligation to act in the public interest, then there is less reason to challenge consent and little reason to regard participation as actually or potentially exploitative. We don't usually say "are you quite sure you want to?" when people fulfill their moral and civic obligations. We don't usually insist on informed consent in such cases; we are usually content that they *merely* consent or simply acquiesce. When, for example, I am called for jury service, no one says, "only attend if you fully understand the role of trial by jury, due process, etc., in our constitution and the civil liberties that fair trials guarantee."[35]

We must weigh carefully and compassionately what it is reasonable to put to potential participants in a trial for their free and unfettered consideration. However, provided potential research subjects are given adequate information,[36] and are free to participate or not as they choose, then the only remaining question is whether it is reasonable to permit people freely to choose to participate given the risks and the sorts of likely gains. Is it reasonable to ask people to run whatever degree of risk is involved, to put up with the inconvenience and intrusion of the study, and so on, in all the circumstances of the case? These circumstances will include both the benefits to them personally of participating in the study and the benefits that will flow from the study to other persons, persons who are of course equally entitled to our concern, respect, and protection (if any are). Putting the question in this way makes it clear that the standards of care and levels of protection to be accorded to research subjects who have full information must be, to a certain extent, study relative.

It is crucial that the powerful moral reasons for conducting science research are not drowned by the powerful reasons we have for protecting research subjects. There is a balance to be struck here, but it is not a balance that must always and inevitably be loaded in favor of the protection of research subjects. They are entitled to our concern, respect, and protection to be sure, but they are no more entitled to it than are, say, the people whom, for example, HIV/AIDS or other major diseases are threatening and killing on a daily basis.[37]

It is surely unethical to stand by and watch 3 million people die this or any year of AIDS[38] and avoid taking steps to prevent this level of loss—steps which will not put lives at risk and which are taken only with the fully informed consent of those who participate. Fully informed consent is the best guarantor of the interests of research subjects. While consent is not foolproof, residual dangers must be balanced against the dangers of not conducting the trial or the research, which include the massive loss of life that possibly preventable diseases cause.[39]

An interesting limiting case is that in which the risks to research subjects are significant and the burdens onerous, but where the benefits to other people are equally significant and large. In such a case the research is both urgent and moral but conscription would almost certainly not be appropriate because of the unfairness of conscripting any particular individual to bear such burdens in the public interest. That is not of course to say that individuals should not be willing to bear such burdens, nor that it is not their moral duty so to do. In fact, the history of science research is full of examples of people willing to bear significant risks in such circumstances; very often these have been the researchers themselves.[40]

Benefit Sharing

I have so far said nothing about the public–private divide in research funding and about the fact that much of the research we have referred to has been carried out in the private sector for profit. This has inevitably led to a concentration both on what the comedian Tom Lehrer[41] memorably called "diseases of the rich" and on conditions where, for whatever reason, a maximum return on investment is to be expected. Here we simply note that the duty to participate in research is not a duty to enable industry to profit from moral commitment or basic decency, and that fairness and benefit sharing as well as the widest and fairest possible availability of the products of research is, as we have seen, an essential part of the moral force of the arguments for the obligation to pursue research. Benefit sharing must therefore be part of any mechanisms for implementing an obligation to participate in research.

A New Principle of Research Ethics

A new principle of research ethics suggests itself as an appropriate addition to the Declaration of Helsinki:

> Biomedical research involving human subjects cannot legitimately be neglected, and is therefore both permissible and mandatory, where the importance of the objective is great and the risks to and the possibility of exploitation of fully informed and consenting subjects is small.[42]

Thus, while fully informed consent and the continuing provision to research subjects of relevant information does not eliminate all possibility of exploitation,[43] it does reduce it to the point at which it could no longer be ethical to neglect the claims and the interests of those who may benefit from the research. It should be noted that it is fully informed consent, and the concern and respect for the individual that it signals, which severs all connection with the Nazi experiments and the concerns of Nuremberg, and which rebuts spurious comparisons with the Tuskegee study.[44] It is this recognition of the obligation to show equal concern and respect for all persons, which is the defining characteristic of justice.[45] The recognition that the obligation to do justice applies not only to research subjects but also to those who will benefit from the research must constitute an advance in thinking about international standards of research ethics.

But on Whom Does the Obligation to Participate in Research Fall?

The Declaration of Helsinki states:

> Medical research is only justified if there is a reasonable likelihood that the populations in which the research is carried out stand to benefit from the results of the research.[46]

Me and My Kind

It is sometimes claimed that where consent is problematic or, as perhaps with genetic research on archival material, where the sources of the material are either dead or cannot be traced, that research may be legitimate if it is for the benefit of the health needs of the subjects or of people with similar or related disorders.[47] The suggestion that research which is not directly beneficial to the patient be confined to research that will benefit the category of patients to which the subject belongs seems not only untenable but also offensive. What arguments sustain the idea that the most appropriate reference group is that of fellow sufferers from a particular disease, Alzheimer's, for example? Surely any moral obligation I have to accept risk or harm for the benefit of others is not plausibly confined to those others who are narrowly like me. This is surely close to claiming that research should be confined to others who are "black like me" or "English like me" or "God-fearing like me"? The most appropriate category is surely "a person like me."[48]

Children and the Incompetent

But what about children?[49] Do they have an obligation to participate in research and, if they have, is a parent justified in taking it into account in making decisions for the child?

If children are moral agents, and most of them (except very young infants) are, then they have both obligations and rights, and it will be difficult to find any obligations that are more basic than the obligation to help others in need. There is therefore little doubt that children share the obligation argued for in this chapter: to participate in medical research. A parent or guardian is accordingly obliged to take this obligation into account when deciding on behalf of her child and is justified in assuming that the person they are making decisions for is, or would wish to be, a moral person who wants to or is in any event obliged to discharge his or her moral duties. If anything is presumed about what children would have wished to do in such circumstances, the presumption should surely be that they would have wished to behave decently and would not have wished to be free riders. If we simply consult their best interests (absent the possibility of a valid consent), then again, as this chapter has shown, participation in research is, other things being equal, in their best interests. However, because of the primacy of autonomy in the structure of this argument we should be cautious about enrolling those who cannot consent in research and should never force resisting incompetent individuals to participate. It also follows from principles of justice and fairness that those who are not competent to consent should not be exploited as prime candidates for research. We should always therefore prefer autonomous candidates and only use those who cannot consent when such individuals are essential for the particular research contemplated and where competent individuals cannot, because of the nature of the research,[50] be used. In those extreme cases in which we might contemplate mandatory participation, the same will hold. The incompetent should only be used where competent individuals cannot be research subjects because of the nature of the research itself.

Inducements to Participate in Research

Before concluding, a word needs to be said about inducements to research, not least because inducements are an obvious alternative to mandatory participation in research. Most research ethics protocols and guidelines are antipathetic to inducements. For example, the CIOMS guidelines state that if inducements to subjects are offered, "[t]he payments should not be so large, however, or the medical services so extensive as to induce prospective subjects to consent to participate in the research against their better judgment (undue inducement)."[51]

However, the gloss that the CIOMS document offers on this guideline is perhaps confused. It states that "[s]omeone without access to medical care may or may not be unduly influenced to participate in research simply to receive such care."[52] The nub of the problem is the question: what is it that makes inducement *undue*? If inducement is undue when it undermines "better judgment," then it cannot simply be the level of the inducement nor the fact that it is the inducement that makes the difference between participation and nonparticipation that undermines better judgment. If this were so, all jobs with attractive remuneration packages would constitute "undue" interference with the liberties of subjects and anyone who used their better judgment to decide whether a total remuneration package plus job was attractive would have been unduly influenced.[53]

Surely, it's only if things are very different that influence becomes undue. If, for example, it were true that no sane person would participate in the study and only incentives would induce them to disregard "better judgment" or "rationality," or if the study were somehow immoral, or participation was grossly undignified and so on, would there be a legitimate presumption of undue influence.

Grant a number of assumptions: research is well-founded scientifically and has important objectives which will advance knowledge; the subjects are at acceptable levels of risk given the benefits;[54] the inconvenience, and so on, of participation is not onerous. Then surely it is not only in everyone's best interests that *some* people participate but also in the interests of those who do. *Better judgment* surely will not indicate that any particular person should not participate. Of course, someone consulting personal interest and convenience might not participate: "it's too much trouble, not worth the effort, rather inconvenient," and so on. However, removing the force of *these sorts of objections* with incentives is not undermining *better judgment* any more than is making employment attractive.[55]

Of course, inducements may be undue in a different sense, for example, if a research subject were a drug addict and she were to be offered the drug of her choice to participate, or subjects were blackmailed into participating in research; in such cases we might regard the inducements as undue. However, it is important to note that here the influence or inducement is undue not because it is improper to offer incentives to participate, nor because participation is against the best interests of the subject, nor because the inducements are coercive in the sense that they are irresistible, but rather because the *type of incentive* offered is illegitimate or against the public interest or immoral in itself.

If I offer you a million dollars to do something involving minimal risk and inconvenience, something that is good in itself, is in your interests, and will benefit mankind, my offer may be irresistible but it won't be coercive. However, if I threaten you with torture unless you do the same thing, my act will be coercive even if you were going to do it whether or not I threatened you. I should be punished for my threat or blackmail or criminal offer of illegal substances, but surely you should nonetheless do the deed and your freedom to do it should not be curtailed because of my wrongdoing in attempting to force your hand in a particular way. The wrong is not that I attempted to force your hand, but resides rather in the wrongness of the methods that I chose. This is the distinction between undue inducement and inducements which are undue. "Undue inducement" is the improper offering of inducements, improper because no inducements should be offered. It is this that is referred to in the various international protocols we have been examining and which is almost always wrongly understood and wrongly applied. "Inducements which are undue," refer to the nature of the inducement, not to the fact of it being offered at all. This is an important but much neglected distinction. Here it is the nature of the inducement that is undue rather than the fact of inducements of some sorts (even irresistible sorts) being offered.

We can see that offering incentives, perhaps in the form of direct payment or tax concessions to people to participate in research, or, for example, to make archive samples available for research, would not be unethical.

One of the worries about payment for research participation is that it will "crowd out willingness to participate" by suggesting that participation is bad for you and requires "compensation." I have frequently suggested that neither compensation nor reward undermine altruism; rather, they complement it. Nursing and medicine are caring professions despite the fact that the "professionals" are rewarded for their altruistic choice of vocation. Indeed, there could scarcely be an altruistic vocation that was not rewarded financially as well as morally. If a possible further response were needed (which it is not), it is that, as Julian Le Grand's discussion of Richard Titmuss reveals, strong *enough* egoistic incentives (based, for example, on the prospect of particularly high remuneration for contribution), while admittedly diminishing altruistic tendencies, usually generate very strong prudential reasons to contribute. The prudential motivation that they generate can more than compensate for the motivational loss in altruistic terms.[56]

Conclusion

There is then a moral obligation to participate in medical and science research more generally in certain contexts.[57] This will obviously include minimally invasive and minimally risky procedures such as participation in Biobanks provided safeguards against wrongful use are in place. The argument concerning the obligation to participate in research should be compelling for anyone who believes that there is a moral obligation to help others, and/or a moral obligation to be just and do one's share. Enhancements being both necessarily good and often very significant goods indeed share importantly in this general obligation. Little can be said to those whose morality is so impoverished that they do not accept either of these two obligations.

Furthermore, we are justified in assuming that a person would want to discharge his or her moral obligations in cases in which we have no knowledge about their actual preferences. This is a way of recognizing them as moral agents. To do otherwise would be to impute moral turpitude as a default. Parents making decisions for their children, and other surrogate decision makers, are therefore fully justified in assuming that their child, or the individual for whom they are empowered to decide, will wish to do that which is right, and not do that which is wrong. Research then is a necessary part of enhancement; it requires commensurate support and endorsement.

It is appropriate that we end looking to the future, reminding ourselves of the giants on whose shoulders we stand and defending the blue skies research that has, since Plato's thought experiments and Aristotle's observations,[58] been not only the inspiration of science and of intellectual enterprise, but the birth, basis, and backbone of the life of the mind.

Notes

Introduction

1. Marx (1972). What Marx actually said was: "The philosophers have only interpreted the world, in various ways; the point, however, is to change it."
2. Russell (2005, p. 91).
3. This I have examined at length elsewhere (see Harris 1985).

1 | Has Humankind a Future?

1. See Harris (1992, 2000a).
2. An extended argument for these assertions was given in Harris (1980).
3. Harris (2001b).
4. Russell (1961, pp. 13–14).
5. The quotations from Russell might be taken as an exercise in citing authorities in support of my position. Never! The only authority is compelling evidence and argument. In citing Russell and indeed Tomasi de Lampedusa I am not appealing to authority but attempting to place myself in a tradition of thought and reflection on the nature and desirability of change, on the balance between conservatism and revolution.
6. Russell (1961, p. 121).
7. Otherwise we are like the ape in the example that follows. Here I am indebted to Nir Eyal for helping me to clarify my ideas.
8. Lampedusa (2005).
9. The Prince of Salina remembers with approval Tancredi's remark (Lampedusa 2005, p. 28) and later votes for change in the famous plebiscite, and his acceptance of Tancredi's wisdom is revealed not only by his wistful "if we want things to stay as they are…" but by his acknowledgment that

"Tancredi would go far: he'd always thought so" (Lampedusa 2005, p. 28) and also by his priest, Father Pirrone, who, when asked what the "prince of Salina feels about the revolution," replies "I can tell you that at once and in a few words; he says there's been no revolution and that all will go on as it did before" (Lampedusa 2005, p. 161).

10. James Hughes has made this point strongly (personal communication, March 15, 2006).
11. Brutus in Shakespeare's *Julius Caesar* justifies the murder of Caesar by reasoning that although Caesar is only a potential and not an actual danger, the prevention of the danger is justified: "Fashion it thus; that what he is, augmented / Would run to these and these extremities; / And therefore think him as a serpent's egg / Which, hatch'd, would, as his kind, grow mischievous, / And kill him in the shell" (act 2, scene 1).
12. See the table "Adult and youth literacy rates," at http://portal.unesco.org/education/en/ev.php-URL_ID=41637&URL_DO=DO_TOPIC&URL_SECTION=201.html. For a table of data concerning "Estimates and projections," see www.uis.unesco.org/TEMPLATE/html/Exceltables/education/View_Table_Literacy_Country_Age15-24.xls; see also http://www2.unesco.org/wef/en-leadup/findings_excluded%202.shtm for a chapter called "Patterns of exclusions: causes and conditions," from "Education for all and children who are excluded," by Anne Bernard, coordinated by the United Nations Children's Fund. The main site for the World Education Forum is http://www2.unesco.org/wef/en-leadup/findings.shtm. (All sites were accessed on May 30, 2005.)
13. In 2003 a paper published by a group from University College London (Maguire et al. 2003) showed that the volume of gray matter in the posterior hippocampus of London taxi drivers increases due to the navigational expertise and knowledge that they acquire through their jobs.
14. Dawkins (2004, pp. 23–31).
15. See, for example, Fukuyama (2002), Sandel (2004), and Annas's "Genism, racism and the prospect of genetic genocide" (www.thehumanfuture.org/commentaries/annas_genism.html; accessed August 11, 2006; link now discontinued; updated version available at www.thehumanfuture.org/commentaries).
16. Harris (1992).
17. Much, of course, remains impossible.
18. Russell (2005, p. 1).

2 | Enhancement Is a Moral Duty

1. I first used this analogy at a public meeting in Cheltenham Town Hall on November 18, 2005. Steven Rose (2006) has recently used the same idea in his "Brain gain?".
2. Boorse (1981).

3. Daniels (1996, p. 185).
4. Rather than to make an excuse for the order he was continuing to disobey (or perhaps to claim priority in the debate about enhancement!).
5. Rose (2006, p. 74).
6. For some verses (if not chapters) see the collection of quotations, allegedly from those "Calling for Bans on Species Altering Technologies" assembled at http://www.genetics-and-society.org/overview/quotes/opponents.html (accessed August 11, 2006). I am grateful to Dr. Sarah Chan for pointing this out to me and for the source of Annas's other excursions into this debate.
7. See Baltimore (2003) for a conference presentation and his "Engineering immunity: a proposal," at www.fastercures.org/printable.php?page_name=essay_baltimore (accessed March 1, 2006).
8. Fukuyama (2002, p. 160).
9. Ibid., p. 101.
10. Ibid., p. 149.
11. Glover (2006).
12. I am indebted to Nir Eyal for this suggestion.
13. "Genism, racism and the prospect of genetic genocide" (for details see chapter 1, note 15).
14. Ibid.
15. Indeed, elsewhere, Annas, along with Lori B. Andrews and Rosario M. Isasi, calls for an international treaty to ban the use of such technologies (see Annas et al. 2002). For the record I have dealt with Annas-style objections to cloning elsewhere (see Harris 2004a).
16. See Harris (2004, p. 50).
17. Personal communication, March 25, 2007.
18. See Harris (1980, p. 50). I say there and still maintain that "I think the conclusion to be drawn is that 'intention,' which lends itself to such sophistical arguments, is not much help in determining moral responsibility or in distinguishing the moral quality of different actions with the same consequences. It looks as though intention can be so narrowly defined as to yield any moral answer that is wanted."
19. I pointed this out in Harris (1980).
20. For more on double effect see Harris (2000b).
21. Jones et al. (2005).
22. Rose (2006).
23. See, for example, Savulescu et al. (2004).
24. I believe this despite many powerful arguments to the contrary. For more on the moral permissibility and impermissibility of advantage in education see Swift (2003, 2004a,b) and Brighouse (1998, 2000a,b, 2002, 2005). For a critique of Brighouse's arguments see Foster (2002) and Tooley (2003).

25. Statistics from Eurotransplant seem to suggest that waiting lists *do* still exist in Belgium and Austria (Eurotransplant does not extend to Spain), although they may be greatly reduced.
26. The resulting loss of life is of course due to a number of factors including the shortage of donor organs, lack of dialysis facilities and material resources, etc.
27. See Harris (1980).
28. For one example of the necessity or at least desirability of such a system see Erin and Harris (1994).
29. Resource allocation is something I have worked on in the past, and further work will result in a book entitled *The Safety of the People* (forthcoming). For further thoughts on these issues see Harris (1987, 1996, 1999) and Erin and Harris (1994).
30. Porter (1999).
31. See Bodnar et al. (1998) and also Weinrich et al. (1997) and McBrearty et al. (1998). For a comprehensive roundup of related work see de Grey (2004b).
32. See Harris (2000c, 2002, 2004b).
33. Søren Holm and I discussed this principle in Harris and Holm (2002), and I draw on this reference where we detail the problems to which I refer in this section.
34. UNESCO Press Release no. 97-29; see also UNESCO "Universal Declaration on the human genome and human rights," December 3, 1997.
35. While some maintain that human evolution of the Darwinian sort is at an end because most humans now survive long enough to reproduce, this view overlooks the role of parasites in evolution. It also of course ignores our deliberate interventions in the evolutionary process.
36. Cornford (1908).
37. Harris (1985, p. 38).
38. Hobbes (1960, p. 82). Julian Savulescu and I made similar points in our paper "The creation lottery: final lessons from natural reproduction: why those who accept natural reproduction should accept cloning and other Frankenstein reproductive technologies" (Savulescu and Harris 2004).

3 | What Enhancements Are and Why They Matter

1. Literally to "raise up." As I use the term it means anything that, nondetrimentally, raises the powers or capacities of humans. See *The Shorter Oxford English Dictionary*, revised and edited by T. E. Onions, 3rd edn (Oxford University Press, 1965).
2. Harris (1992).
3. Daniels (2007).
4. Buchanan et al. (2000, chapter 3).
5. See Harris (1992, chapters 7–9).
6. I am indebted to Nir Eyal for his comments here.

7. Although of course, since they would, being my children, have had a sense of humor, they wouldn't have been laughing.
8. UNESCO Press Release no. 97-29; see also UNESCO "Universal Declaration on the human genome and human rights," December 3, 1997.
9. Daniels (2007; see below).
10. Ibid.
11. Harris (1992); I quote from chapter 9.
12. Ibid., pp. 201–2.
13. Boorse (1981).
14. Daniels (1996, p. 185).
15. Harris (2001b).
16. See chapter 4. This remark was made by Tom Kirkwood (see Harris 2000c, 2004c).
17. Daniels (2000, p. 313).
18. Buchanan et al. (2000, p. 71).
19. See Dworkin (1981a,b), both reprinted in Dworkin's *Sovereign Virtue*, pp. 11–120 (Harvard University Press, 2000), and Scanlon (1989).
20. Buchanan et al. (2000, p. 74).
21. Ibid., p. 73.
22. Harris (1980, 1985).
23. Buchanan et al. (2000, p. 74).
24. This way of defining health and illness is derived from Boorse (1981) but is used also by Daniels (1996, p. 185), and many others.
25. Buchanan et al. (2000, p. 75).
26. The obsession with equal opportunities exemplified by these and other writers on health seems to be an unhealthy legacy of Rawls and the prestige of theories of (or even theories mentioning) justice. Justice is of course part of ethics but some write and appear to think that it is all of ethics.
27. "The office of the sovereign, be it a monarch or an assembly, consisteth in the end for which he was trusted with the sovereign power, namely the procuration of the safety of the people; to which he is obliged by the law of nature" (Hobbes 1960).
28. Daniels (2007).
29. Harris (1999a, 2005c).
30. See World Medical Association (1964), with the note of clarification of paragraph 29 added by the WMA General Assembly, Washington, 2002.
31. See chapter 11, below.
32. This is the view of Aubrey de Grey (personal communication, March 21, 2005). See also de Grey (2004a), in which he speculated that "the first 1,000-year-old is probably only 5–10 years younger than the first 150-year-old."
33. To those to whom it is acceptable.
34. I discuss these issues in the next chapter but also in Harris (2000d) and the similarly titled but fuller treatment (Harris 2002a).

35. This is indeed how I would characterize it.
36. Some indications as to how I would address these issues are to be found, inter alia, in Harris (1980, 2005d) and Harris and Sulston (2004).
37. See footnote 1 of Daniels (2007).
38. Daniels (2007).
39. Perhaps so that enhancements are easier targets for puerile objections?

4 | Immortality

1. This chapter draws on Harris (2000d, 2004b).
2. Adams (1982, p. 9). For the record, the immortal's name was "Wowbagger."
3. And we should note that Wowbagger himself did find something meaningful to do through all eternity.
4. See, for example, Weiss (1985).
5. I am grateful to Simon Woods for insights into the undead.
6. See, for example, Slevin et al. (1990).
7. I have benefited from the incisive comments of my colleague Søren Holm.
8. Bodnar et al. (1998), Weinrich et al. (1997).
9. McBrearty et al. (1998).
10. See Thomson et al. (1998), Pedersen (1999), Mooney and Mikos (1999).
11. Lanza et al. (1999a,b).
12. These possibilities were rehearsed in the BBC TV *Horizon* program "Life and Death in the 21st Century," broadcast in January 2000.
13. For a detailed defense of this idea see Harris (1985).
14. This claim is defended in detail in Harris (1980, 1985).
15. I use the terms "immortals" and "mortals" as shorthand for those who do and do not benefit from significantly increased life expectancy; see also Silver (1998).
16. See J. Carvel "North–south life expectancy gap grows wider," Guardian Unlimited, October 16, 2004 (available at www.guardian.co.uk / north south / article / 0,2763,1328652,00.html; accessed March 12, 2006).
17. According to the U.S. Census Bureau's International Database, quoted at http: / / geography.about.com / library / weekly / aa042000b.htm, accessed March 12, 2006.
18. Harris (2002a).
19. See Jonas (1992), Glannon (2002).
20. Kass (2001).
21. Ibid.
22. This, incidentally, is a problem for all accounts of disability that see persons with disabilities as simply "differently abled" (see Harris 2000a).
23. Glannon (2002).
24. Personal communication, April 8, 2000; see also Austad (1997). Austad has also calculated that at this constant death rate about one person in a

thousand would live to be 10,000 years old, which is pretty close to the rate at which people live to be 100 years old today. There may be problems with these calculations, however. Søren Holm has pointed out to me that eleven- or twelve-year-olds may not be at the normal risk for accidents, being somewhat protected by their parents and unlikely to be drivers of motor vehicles. These calculations are also almost certainly only good for high-income countries.

25. S. Austad. Personal communication, October 2006.
26. There is an enormous literature on this; see, for example, Harris (1987, 1997b) and McKie et al. (1998, p. 151).
27. See Harris (1998, pp. 5–37; 1999d, pp. 61–95).
28. I deliberately choose the term "generational cleansing" for its obvious unpalatable connotations.
29. Christine Overall, in her recent book (Overall 2005), has found it difficult to be sure of my attitude toward "generational cleansing." For the record, I think it would be unjustifiable and therefore it is difficult to see how we could resist death-postponing therapies.
30. Personal communication, April 1, 2004; the calculations are his.
31. Douglas Adams used a similar argument to show that the cost of traveling in time to eat at "the restaurant at the end of the universe" would bring the price of eating at the most expensive restaurant of all time easily within reach of the most humble budget: "All you have to do is deposit one penny in a savings account in your own era, and when you arrive at the End of Time the operation of compound interest means that the fabulous cost of your meal has been paid for" (see Adams 1980, p. 81).

5 | Reproductive Choice and the Democratic Presumption

1. I am assuming, but not arguing for, the idea that only liberal democracies are worthy of the name and that, as I do suggest here at some length, the tyranny of the majority, which some call majoritarian rule, is not democratic or in any sense worth admiring or fighting for.
2. Mill (1962).
3. Feinberg (1984, pp. 9–11).
4. Dworkin (1977, p. 267).
5. Dworkin (1993).
6. Savulescu (2002a).
7. See Harris (1998, pp. 5–37; 2003a, 2005a).
8. Pedain (2005).
9. See Dworkin (1977, note 31, p. 60).
10. For an eloquent argument for the extension of basic rights in new circumstances see *State of Washington et al., Petitioners, v Harold Glucksberg et al. Respondents,* 117 Sup. Ct. 2258, 138 L.Ed.2d 772 (1997); *Dennis C. Vacco, Attorney General of New York, et al., Petitioners, v Timothy E. Quill,*

et al., Respondents, 117 Sup. Ct. 2293, 138 Nos 5-1858, 96-110 (December 10, 1996). Brief for Ronald Dworkin, Thomas Nagel, Robert Nozic, John Rawls, Thomas Scanlon, and Judith Jarvis Thomson as Amici Curiae in support of respondents.

11. Dworkin (1996, pp. 237–38).
12. Dworkin has produced an elegant account of the way the price we should be willing to pay for freedom may or may not be traded off against the costs; see Dworkin (1977, chapter 10; 1985, chapter 17).
13. See (Dworkin 1996, note 30, pp. 166–67).
14. See Harris (1992, pp. 177ff).
15. See, for example, Harris (1980), Glover (1977), and Rachels (1975, 1979).
16. Despite the best efforts of many, including Frances Kamm (see Kamm 1983, 1992).
17. While it is true that the doctrine of acts and omissions need not say that every harm-doing is worse than similar harm-allowing, it is enough that some harm-doing is worse than similar harm-allowing simply because it involves an act rather than an omission. I have seen no persuasive such examples.
18. Here the argument follows lines set out in Harris (2006).
19. Indeed, doctors at Falkirk and District Royal Infirmary were recently much criticized for so doing (see http://news.bbc.co.uk/hi/english/uk/scotland/newsid_625000/625680.stm).
20. I argued against the relevance of the moral distinction between acts and omissions in Harris (1980). This irrelevance has belatedly been recognized by the highest court in the United Kingdom (see Lord Mustill's judgment in *Airedale NHS Trust v Bland* [1993] 1 All England Rep. 821 H.L.).
21. Of course, I am not here endorsing the spurious and thoroughly discredited claims that MMR is linked to autism. But there are remote and very slight risks attached to all medical interventions, including the very safe MMR vaccine.
22. These ideas are further elaborated in Harris and Keywood (2001).
23. Clarke and Flinter (1996).
24. In the recent case of *Re C (HIV test)* [1999] 2 FLR 1004, it is noteworthy that the High Court did not consider the child's autonomy interests as part of its determination of baby C's best interests. Compelling C's parents to have the child tested for HIV undoubtedly removes the possibility of C deciding at a later date whether to have the test or not, an argument which C's mother put to the court and which was regarded as a "hopeless programme for the baby's protection."
25. There are also arguments which assert and defend a right or entitlement to reproductive liberty to the same effect (Harris 1999d, pp. 61–95; Robertson 1994).
26. Laurie (1999, p. 122).

6 | Disability and Super-Ability

1. This is not of course to say anything about the degree of plausibility.
2. I am of course aware that there are many issues of political correctness concerning the use of terms like "disabled" or "impaired." While I try to use language that in no way could be considered offensive to any group or individual (indeed I was one of the first writers to randomize the gender of the personal pronoun in my work), I am not persuaded that normal use of English idioms in the present context is in any way prejudicial. I therefore continue, for the sake of variety, to move freely between expressions like "people with disabilities" and "the disabled."
3. See also Harris and Sulston (2004).
4. Compare the following questions. Is it wrong to prefer to be a woman? Is it wrong to prefer to be a Jew? Is it wrong to prefer to be black?
5. Or for that matter, to wish it for oneself or one's friends?
6. I discussed this phenomenon in, inter alia, Harris (2000a); see also Spriggs (2002) and Savulescu (2002a).
7. This is of course Derek Parfit's "nonidentity problem" (see Parfit 1984, part 4, especially pp. 358ff). I discuss this problem in the context of disability in Harris (1992, chapter 3). See also Parfit (1976).
8. This is because insertion of two or three embryos maximizes the chances of one successful pregnancy and hence one live birth, and inserting no more than two minimizes the chances of multiple pregnancy which might decrease the chances of getting even one live birth or increase the chances of complications for the mother or the resulting children.
9. It is just this point about the moral importance of the embryo about which those who accept and those who reject abortion disagree.
10. Harris (1992, chapter 4; 1993).
11. See note 10.
12. I have noted this already in discussing enhancements in previous chapters but another example won't hurt.
13. Glover (2006).
14. Ibid., p. 25.
15. And hence have no reason to forbear the exercise of any right to reproduce.
16. See Harris (1998, pp. 5–38).
17. Harris (1992).
18. When I talk of valuing individuals in an existential sense, valuing an individual "as a person," I am talking of the dimension of the individual which entitles them to equal concern, respect, and protection, which speaks to the issue of their dignity and standing as citizens with equal moral and political claims. I have analyzed this notion at some length in Harris (1995, chapter 1).
19. Davis (1988, p. 150). See also Davis (1994) and Harris (1994).
20. See Harris (1985, chapter 1; 1994).

21. I developed the parallel between attempting to eradicate disability and ordinary medical care in, inter alia, Harris (1993).
22. See note 23, below.
23. There is Judith Thomson's famous article (Thomson 1971) but it is unpersuasive on the subject of self-defense because it does not adequately show why the fetus is not entitled to defend itself (and hence be defended by others) as vigorously as the mother.
24. Unattractive particularly to women, though of course not indefensible on that ground alone. There are other grounds for rejecting it, however. See Harris (1985, chapter 1; 1999b; 2002d).
25. Locke (1964, book II, chapter 27).
26. And I have given one such account on a number of occasions (see Harris 1985, 1992).
27. I personally find the views of anyone who would sacrifice the life of a mother to save the life of a fetus simply repugnant and more so when the fetus will be unlikely to survive to term let alone long after birth. However, I would never consider denying such a person the right to express such views.
28. A rather different point often confused with the alternative formulation.
29. The damage must of course leave me with sufficient brain function for self-consciousness.
30. And even animals for that matter.
31. Although of course even embarrassments of this magnitude do not of themselves demonstrate the error of a position to which they attach.
32. I use Jonathan Glover's phrase "ugly attitudes" (see below).
33. Glover (2006, pp. 29ff).
34. Ibid., p. 35.
35. See Koch (2001) and Edwards (2001).
36. Koch (2001, p. 372).
37. Edwards (2001).
38. Cited in Albert (2001).
39. Koch (2001, p. 374).
40. See Harris (1992, pp. 179ff).
41. The counterfactuals here are very problematic. Suppose Hitler had never been born—would that have necessarily meant that the world would have been saved from the Holocaust?
42. See Harris (1985, 1999b).
43. I ignore problems about the scope of the personal pronoun here because, insofar as they are problems, they support my argument.

7 | Perfection and the Blue Guitar

1. In writing this chapter I have been helped by conversations with Dr. Daniela Cutas and Dr. Katrien Devolder.

2. Sandel (2004).
3. Confucius, born in 551 B.C. in the state of Lu, now Shantung, in eastern China had much to say about the role of medicine in the world (see Tsai 2005; Lee 1994). His near(-ish) contemporary Hippocrates (born on Kos in 460 B.C.) is famous for considering related themes. 551 B.C. to 451 B.C. was a good hundred years for bioethics (see also Eliot 1909–1914).
4. W. S. Gilbert, *The Gondoliers*, act 2 (see Gilbert 1956, p. 543). Andy Warhol's often misquoted (including by himself) "In the future everyone will be famous for 15 minutes" expresses a related sentiment. The expression is, according to Wikipedia, a paraphrase of Andy Warhol's statement in 1968 that "In the future, everyone will be world-famous for 15 minutes." In 1979, Warhol reiterated his claim: "...my prediction from the sixties finally came true: [i]n the future everyone will be famous for fifteen minutes" (see http://en.wikipedia.org/wiki/15_minutes_of_fame#_note-0#_note-0; accessed December 19, 2006).
5. Ignoring one of the most coherent and compelling formulations of the precautionary principle "when you're in a hole, stop digging."
6. "And I looked, and behold a pale horse: and his name that sat on him was Death, and Hell followed with him. And power was given unto them over the fourth part of the earth, to kill with sword, and with hunger, and with death, and with the beasts of the earth" (Revelation 6:8, not Keith Reid 1967).
7. A discussion of "closure" in another usage is consigned to the devotees of psychobabble.
8. Those familiar with Uriah Heep will find it difficult to be over fond of humility.
9. In times past cricket might have been an exception, with the likes of two famous exponents of the sport: Colin Cowdrey, not noticeably fit, and Dennis Compton, not noticeably hard working.
10. The phrase "effortless superiority," which students of Balliol College, Oxford, like to associate with themselves, is usually attributed to Herbert Asquith, who apparently referred to Balliol men (they were only men in those days) as having "the tranquil consciousness of effortless superiority" (see Jones 2005, chapter 16). Anthony Kenny, a recent Master of Balliol, confided to me that unfortunately the undergraduates in his time "found the effortlessness much easier than the superiority." A further, if anecdotal, reminder that even those apparently endowed with the "gift" of a Balliol education cannot altogether dispense with effort.
11. A much quoted proverb but I have been unable to find an authentic source.
12. Sandel (2004).
13. Ibid.

14. "'When *I* use a word,' Humpty Dumpty said, in a rather scornful tone, 'it means just what I choose it to mean, neither more nor less'" (Carroll 1940, p. 196).
15. See Harris (2005a).
16. I have written about this in many places. See Harris (1980, 1992).
17. Adams (1980).
18. Ibid., chapter 5, p. 28.
19. Which might include being eaten alive by the Ravenous Bugblatter Beast of Traal (see Adams 1979, p. 45).
20. See Harris (1980).
21. See Cohen (1989, 2000) and Temkin (1993, 2003, 2004).
22. I am indebted to Nir Eyal for this point.
23. Carroll (1940, p. 196).

8 | Good and Bad Uses of Technology: Leon Kass and Jürgen Habermas

1. Glover (2006, pp. 83–85).
2. See Kass (2003). This paper reflects views Kass presented to The United States President's Council on Bioethics in January 2003 (see http://bioethicsprint.bioethics.gov/background/kasspaper.html).
3. Kass (2003, p. 13).
4. Ibid., p. 16.
5. Harris (2004a).
6. Ibid., chapter II.
7. Kass (2003, pp. 18–19).
8. I do not of course endorse the so-called "test of time," which has given us millennia of despotism, racism, and sexism, not to mention other cultural and religious prejudices against strangers, apostates, and heretics.
9. Boorse (1981).
10. Daniels (2007).
11. I have defended postmenopausal mums from such attacks elsewhere (see Harris 1998, pp. 5–37).
12. Holm (1998).
13. Kass (1997, p. 23).
14. Harris (2004a, chapter 3).
15. Tolstoy (1945, p. 13).
16. Kass (2003, p. 17).
17. Kass (1997).
18. Harris (2004a, chapter 2).
19. See Kass (1997, pp. 17–26). The obvious erudition of his writing leads to expectations that he might have found feelings prompted by more promising parts of his anatomy with which to entertain us.
20. Reliable as sources for the story, not of course for the veridical nature of the story.

21. The Bible is I believe silent on the question of how Eve lost the extra Y chromosome.
22. Dolly was the first mammalian clone (Wilmut et al. 1997; see also text below). Dolly-style clones do not share the mitochondrial DNA which exists in the egg prior to fertilization of nuclear substitution, unless of course the cell nucleus from a woman's adult cell is used to clone with her own egg.
23. In a letter to Humphrey House, April 11, 1940 (Orwell 1970, p. 583). See my more detailed discussion of the problems with this type of reasoning in Harris (1992, chapter 2).
24. Kass (2003, p. 19).
25. Ibid., p. 20.
26. My evidence for this is anecdotal.
27. Ibid., p. 23.
28. Actually, like Kass, and unlike the Balliol man I am, I quite enjoy and appreciate the effort. But that is not to say that it is appropriate to impose these personal and perhaps eccentric preferences on those who believe in the ethic of "laborsaving" devices that made America great.
29. For those who can afford an oceangoing sailboat and think they can manage without a chronometer or satellite navigation.
30. Ibid., p. 26.
31. Ibid., p. 27.
32. Or at least no loss.
33. Glover (1984, chapters 7 and 8).
34. Stoppard (1972, act 1, p. 24).
35. Kass (2003, p. 28).
36. Habermas (2003).
37. Habermas does not reference this quotation and I have been unable to find it in English.
38. Ibid., p. 48.
39. Ibid., p. 66.
40. Ibid., p. 62.
41. Ibid., p. 63.
42. Ibid., p. 67.
43. Ibid., p. 66.
44. I was told this story by that student, but whether he in turn was told the story by a man who danced with a girl who danced with the Prince of Wales I cannot be sure.
45. Ibid., pp. 75ff.
46. Ibid., p. 82.
47. With deference to Ludwig Wittgenstein (1966, paragraph 6.45).
48. Habermas (2003, p. 86).

49. I have argued for the truth of this claim in exhaustive detail elsewhere; see Harris (1980).

9 | Designer Children

1. An example suggested to me by Jonathan Glover (personal communication).
2. For some answers to this question which are at odds with what is said in this book see Scott (2006), Sandel (2002), and The President's Council on Bioethics Report on Enhancement (available at www.bioethics.gov/topics/beyond_index.html).
3. I have examined these questions at length on a previous occasion (see Harris 1992, pp. 158ff).
4. I am here indebted to Nir Eyal (personal communication).
5. See Giordano (2007).
6. See Harris (1998, pp. 5–37).
7. Figures for the United Kingdom in 2004 indicate 1,255,000 more women than men. Statistics from the website of the Equal Opportunities Commission (www.eoc.org.uk/Default.aspx?page=14895).
8. See Human Fertilisation and Embryology Authority (2003).
9. Parts of the discussion that follows were published in Harris (2005b,d). See also the response by Baldwin (2005).
10. D. W. Harding, "Regulated hatred: an aspect of the work of Jane Austen," *Scrutiny* (1940). The version I have used appears in Lodge (1972, p. 262).
11. "He led his regiment from behind— / he found it less exciting. / But when away his regiment ran, / his place was at the fore, O..." W. S. Gilbert (1956, p. 510), *The Gondoliers*.
12. Almost certainly different children but not necessarily so.
13. The *Human Fertilisation and Embryology Act* 1990 (Clause 13.5) states: "A woman shall not be provided with treatment services unless account has been taken of the welfare of the child who may be born as a result of the treatment (including the need of that child for a father), and of any other child who may be affected by the birth." As it happens I think both procedures would maximize child welfare. I simply doubt the HFEA would agree.
14. I have argued this point at length elsewhere; see Harris (2000c) and Burley and Harris (1999).
15. I owe this characterization of the nonidentity arguments to Julian Savulescu, but from the horse's mouth you will find it in Parfit (1984, chapter 16). See also Burley and Harris (1999).
16. Baldwin (2005).
17. Habermas (2003); see also the discussion of Habermas in the previous chapter.

18. Parts of this section were published in the "Podium" column, *The Independent* (London), November 27, 2003, p. 19.
19. Nir Eyal, in his referee report, helped me clarify this point.
20. See Harris (1998, pp. 5–37).

10 | The Irredeemable Paradox of the Embryo

1. This chapter owes much to the work of Katrien Devolder and to the insightful comments of Nir Eyalin in his referee report.
2. See Devolder (2006). Some references to those who have suggested alternative human embryonic stem (hES) cell sources are Hurlbut (2005), Landry and Zucker (2004), Liao (2005), and President's Council on Bioethics (2005).
3. See Aristotle's *De Anima* (book II, no. 1) and *Metaphysics* (book VII, no. 10) in Barnes (1984).
4. For more on this ambiguity see Harris (1983, p. 132; 1997a, 1999b), and Burley and Harris (1999). Recently Dan Brock has also discussed these issues (see Brock 2006).
5. See Harris (1998, pp. 5–37) and the "Introduction" in Harris (2004a).
6. For elaboration of my view of the problematic nature of attributing moral significance to early embryos see Harris (1985, 2003d).
7. See the section on "rights."
8. I discuss the distinction between harming and wronging in Harris (1992, chapter IV).
9. See Cohen (1995, p. 68) (this originally appeared in Cohen (1986, p. 109)) and see also Steiner (1994). On rights theories generally, see Sumner (1987), Waldron (1988, chapter 3), Steiner (1994), and Dworkin (1977; 1993, pp. 210–16).
10. See Raz (1986, p. 166). A related conception in terms of welfare rather than well-being is offered by Sumner (1987, p. 47).
11. See The European Court of Human Rights judgment [Case of *Vo v. France* (application no. 53924/00) Strasbourg, July 8, 2004], and, most recently, [*Evans v. the United Kingdom* (application no. 6339/05) Judgment, Strasbourg, March 7, 2006].
12. Adapted from Harris and Holm (2003b, p. 112–36). I thank Søren Holm for permission to adapt these jointly authored ideas and present them here.
13. Devolder and Ward (2007).
14. This has been the subject of the debate between Don Marquis, Julian Savulescu, and others; see Marquis (1989; 2005, p. 119) and Savulescu (2002b).
15. Finnis (1995a,b).
16. Ibid.
17. Harris (1999b).

18. See Takahashi and Yamanaka (2006), Nayernia et al. (2006), Niwa et al. (2005), and Sheng et al. (2003). See also Devolder and Ward (2007). I am indebted to Katrien Devolder for much of the argument of this paragraph (Devolder and Harris 2007).
19. Li et al. (2005a,b); Nagy et al. (1993).
20. Gerami-Naini et al. (2004).
21. Devolder and Ward (2007).
22. Julian Savulescu has advanced this argument (see Savulescu 2002b).
23. Personal communication, January 23, 2007.
24. "Every year an estimated 7.9 million children—6 percent of total births worldwide—are born with a serious birth defect of genetic or partially genetic origin. Additional hundreds of thousands more are born with serious birth defects of post-conception origin, including maternal exposure to environmental agents (teratogens) such as alcohol, rubella, syphilis and iodine deficiency that can harm a developing fetus" (Christianson et al. 2006).
25. In a conversation that, like most human conversations, was memorable for its content and not for time, place, and date on which it took place.
26. See Boklage (1990) and also Leridon (1977).
27. See note 25.
28. See Green (2001, n. 185). A figure of 70% total embryo loss is confirmed by Macklon et al (2002). Edmonds et al. (1982) give a figure of 61.9% loss before twelve weeks, but since this figure does not include embryo loss before implantation or from miscarriage after twelve weeks, the figure of 80% estimated by Winston may not be an unreasonable estimate. See also Hertig and Rock (1973), Adams et al. (1956), Roberts and Lowe (1975), Bovens (2006), and commentary in the New York Times (www.nytimes.com / 2006 / 06 / 13 / health / 13rhyt.html?adxnnl=1 &adxnnlx=1150199902-j6JDYWmEDy8X0gdLASAmNg; accessed July 1, 2006; available via free subscription).
29. While most of those who oppose embryo research oppose abortion in some circumstances (with many exceptions for pregnancies that are a result of rape), many do not also oppose these methods of contraception.
30. Marquis (2005, p. 119). For a thorough account of the debate between Savulescu and Marquis, see Kuflik (2007).
31. See Savulescu and Harris (2004a,b) and Harris (2004c).
32. Harris (2004a, p. 137); I have slightly changed the quotation from *On Cloning* to sharpen its relevance for the present argument.
33. I have argued some of these points in Harris (2002b, 2003d).
34. Not *any* means of course, but at least one possible means.
35. For further arguments relevant to these issues see Savulescu (2004), Harris (2004c), and Savulescu and Harris (2004a).

36. Maurizio Mori gives the following account (in a personal communication) of the origin of the expression "the embryo, one of us." "The story is the following: it was used by Prof. Francesco D'Agostino, Chair of the Italian National Ethics Committee, when he presented the result of the Report of the Italian National Committee for Bioethics to the media in 1996. He said more precisely that the Committee had stated 'the human embryo is to be treated as one of us.' Of course it was immediately shortened to the more famous form." There is, however, at least one book with such a title (see Concetti 1997). The phrase "the embryo is one of us" is now also associated with Pope John Paul II and is widely used in pro-life discourse (see Berthelet (2003) and Congregation for the Doctrine of the Faith "Instruction on respect for human life in its origin and on the dignity of procreation: replies to certain questions of the day" at www.vatican.va / roman_curia / congregations / cfaith / documents / rc_con_cfaith_doc_ 19870222_respect-for-human-life_en.html).
37. Since on the "embryo is one of us" view the moral imperatives are of high importance and urgency, fertile women have a duty not to be too choosy in their choice of available partners. This is why I earlier described all forms of the potentiality argument for moral value as involving an "exhausting ethic."
38. Harris (2003b, pp. 99–109).
39. Harris (2003d).
40. It is now accepted that if future 9/11-style hijackings take place, governments will have to face the question of whether to use Air Force jets or missiles to shoot down the hijacked planes, killing all the passengers innocent *and* guilty. The same trade-off will apply.
41. For more on this and, in particular, the role of Pope John Paul II, see Wrong (2005), Toynbee (2005), and Hari (2005). For a rational dissenting judgment see O'Neill (2005). See also Harris (2003b).
42. Parfit (1984, p. 206).
43. I am indebted to my colleague Matti Hayry for many suggestions in these sections.
44. Kass (2001).
45. Adams (1980, p. 79).
46. Crichton (1999).
47. See Glover (1977, p. 57).

11 | The Obligation to Pursue and Participate in Research

1. This chapter follows lines developed in Harris (2005c).
2. Brecht (1986, scene 14, pp. 108–9).
3. Sandel (2004).
4. Karl Marx (1972).

5. See Brecht (1986, p. 108). It should hardly be necessary to state (although experience teaches otherwise) that these endorsements of views of Marx himself and Brecht, often characterized as a "Marxist playwright," do not imply acceptance of any other of Marx's views or of any other aspects of Marxist philosophy. I hope that in the future it is not only Marx and Marxists who are prominent in highlighting the social and moral responsibility of the generality of humankind.
6. Benevolent authors should always gift at least one devastating quotation to reviewers in the hope of attracting attention.
7. Harris and Holm (2002).
8. See World Medical Association (1964), with the note of clarification of paragraph 29 added by the WMA General Assembly, Washington 2002.
9. Council for International Organizations of Medical Sciences (2002).
10. See Caplan (1992a) and Glover (1999, part 6).
11. See Angell (1997).
12. Caplan (1992b).
13. Redfern (2001). For a commentary on some of the major issues concerning this case see Harris (2002e).
14. Here the argument is restricted to research projects that are not merely aimed at producing knowledge. Unless an increase in knowledge is a good in itself (a question we will not discuss here), some realistic hope of concrete benefits to persons in the future is necessary for the validity of our arguments.
15. Korn (1998).
16. I set out the arguments for and the basis of this duty in Harris (1980).
17. See Barry (1995, p. 228).
18. Jonas (1972).
19. See Hart (1955) and Rawls (1972).
20. This formulation of the principle derives from Nozick (1974, p. 90).
21. It is perhaps also worth pointing out that there is a separate question about whether this moral obligation should be enforced on those who do not discharge it voluntarily. This is not a question I will discuss here.
22. See, for example, Harris (1997c).
23. There are of course imaginable scenarios which would justify even conscription into highly risky research just as there are imaginable scenarios in which killing the innocent is justifiable. The fact that some unscrupulous individuals or governments are all too willing to overexercise their imagination in this direction, for example, in the "collateral damage" to the innocent that often attends military operations, does not destroy the validity of the point.
24. I owe this formulation of the interest I have in being a moral agent to Søren Holm.

25. See also Harris and Holm (1997).
26. See World Medical Association (1964).
27. Here the argument echoes Harris (1999a).
28. And may also have specific contractual duties to them.
29. For example, in cases of research on young children, mental patients, and others whom it is reasonable to assume may not be adequately competent.
30. See Harris (2002c).
31. I use this term in a nontechnical sense.
32. For use of this principle in a different context see Harris (2003c). Taxation is of course the clearest and commonest example.
33. Those over sixty-five may be excused if they wish.
34. I talk here of minimal risk in the sloppy fashion usual in such contexts. However, "risk" is ambiguous between "degree of danger" and "probability of occurrence of danger." Risk may of course be minimal in either or both of these senses.
35. If these suggestions are broadly acceptable and an obligation to participate in research is established, this may well become one of the ways in which research comes to be funded in the future.
36. There is of course no such thing as full information.
37. Of course, the historical explanation of the Declaration of Helsinki and its concerns lies in the Nuremberg trials and the legacy of Nazi atrocities. However, we are, I believe, in real danger of allowing fear of repeating one set of atrocities lead us into committing other new atrocities.
38. Figures are for 2003, with an estimated 5 million people newly acquiring HIV in that same year. Source: Joint United Nations Programme on HIV/AIDS (www.cdc.gov/hiv/stats.htm).
39. These residual dangers include the difficulties of constructing suitable consent protocols and supervising their administration in rural and isolated communities and in populations which may have low levels of formal education.
40. For one prominent example, that of Barry Marshall's work in which he swallowed *Helicobacter pylori* bacteria, thereby poisoning himself in order to test a bacterial explanation for peptic ulcers (see http://opa.faseb.org/pdf/pylori.pdf and www.vianet.net.au/~bjmrshll/).
41. Introduction to track #6, *An Evening (Wasted) With...* Tom Lehrer (1990, REPRISE/WEA 6199, compact disc).
42. For an earlier version of this principle applied in the context of genetics see Harris (1999a).
43. As Marcia Angell rightly points out (Angell 1997).
44. See note 43.
45. See Dworkin (1977).
46. World Medical Association (1964, paragraph 19).

47. See, for example, Council for International Organizations of Medical Sciences (1993, guideline 6, p. 22).
48. I make a distinction between humans and persons which is not particularly pertinent in this context but explains my choice of terminology. See Harris (1985, chapter 1; 1999b).
49. Harris and Holm (2003a).
50. For example, because the research is into an illness which only affects children or those with a particular condition that affects competence, or where the point of research is, for example, to check whether the illness or the treatment affect children in the way that they affect adults.
51. Council for International Organizations of Medical Sciences (2002, guideline 7).
52. Ibid., pp. 28ff.
53. The CIOMS gloss on their own guideline creates a kind of catch-22 which is surely unreasonable and unwarranted. Wherever the best proven diagnostic and therapeutic methods are guaranteed by a study in a context or for a population who would not normally expect to receive them, this guideline would be broken. CIOMS guideline 4 therefore surely contradicts and violates not only the Declaration of Helsinki but also its own later guideline 14.
54. An interesting test case here is the level of risk which is acceptable for live organ donation and the question of the degree to which incentives affect this (see Erin and Harris 2003).
55. See also Wilkinson and Moore (1997) and McNeill (1997) and the discussion of commercial exploitation in Harris (1992, chapter 6).
56. See Le Grand (2006).
57. This obligation has been partly endorsed by the Hugo Ethics Committee in its "Statement on human genetic databases," published in December 2002. However, like so many statements by august ethics committees, the Hugo statement contains not a single argument to sustain its proposals or conclusions. This chapter provides the missing arguments. For a critique of the operation of national and international ethics committees see the introduction to Harris (2001a, pp. 1–25).
58. Of course, since even before Plato and Aristotle, but they constitute a usefully dramatic opening scene and they combine different aspects—the empirical and the theoretical—of what we now think of as research. This idea of course was used by Raphael in his masterpiece *The School of Athens* in the Stanza della Segnatura in the papal apartments in the Vatican. I have not neglected to note that both theoretical and observational aspects of research were subject to error even in the ancient world.

| Bibliography

Adams, D. 1979. *The Hitchhiker's Guide to the Galaxy*. London: Pan Books.

——. 1980. *The Restaurant at the End of the Universe*. London: Pan Books.

——. 1982. *Life, the Universe and Everything*. London: Pan Books.

Adams, E. C., A. T. Hertig, and J. Rock. 1956. A description of 34 human ova within the first 17 days of development. *The American Journal of Anatomy* 98:435–93.

Albert, B. 2001. Genetics: promising cure or delivering elimination? *Consumer Policy Review* 11:166–71.

Angell, M. 1997. The ethics of clinical research in the third world. *The New England Journal of Medicine* 337:847–49.

Annas, G., L. B. Andrews, and R. Isasi. 2002. Protecting the endangered human: toward an international treaty prohibiting cloning and inheritable alterations. *American Journal of Law and Medicine* 28:151–78.

Austad, S. 1997. *Why We Age: What Science Is Discovering about the Body's Journey Through Life*. Wiley.

Baldwin, T. 2005. Reproductive liberty and elitist contempt: reply to John Harris. *Journal of Medical Ethics* 31:288–91.

Baltimore, D. 2003. Using stem cells as targets for gene therapy approaches. Paper presented at The First International Congress of Stem Cell Research, Singapore.

Barnes, J. (ed.). 1984. *The Complete Works of Aristotle*. Princeton University Press.

Barry, B. 1995. *Justice as Impartiality*. Oxford: Clarendon Press.

Berthelet, J. 2003. Statement by the President of The Canadian Conference of Catholic Bishops on Bill 13: an act respecting assisted human reproduction. (Available at www.cccb.ca/site/content/view/1417/1063/lang,eng/.)

Bodnar, A. G., M. Ouellette, M. Frolkis, S. E. Holt, C. P. Chiu, G. B. Morin, C. B. Harley, J. W. Shay, S. Lichtsteiner, and W. E. Wright. 1998. Extension of life-span by introduction of telomerase into normal human cells. *Science* 279:349–52.

Boklage, C. 1990. Survival probability of human conceptions from fertilization to term. *International Journal of Fertility* 35:75–94.

Boorse, C. 1981. On the distinction between disease and illness. In *Medicine and Moral Philosophy* (ed. M. Cohen, T. Nagel, and T. Scanlon). Princeton University Press.

Bovens, L. 2006. The rhythm method and embryonic death. *Journal of Medical Ethics* 32:355–56.

Brecht, B. 1986. *Life of Galileo* (transl. J. Willett). London: Methuen.

Brighouse, H. 1998. Why should states fund schools? *British Journal of Educational Studies* 46:138–52.

——. 2000a. *School Choice and Social Justice*. Oxford University Press.

——. 2000b. *A Level Playing Field: The Reform of Private Schools*. London: Fabian Society.

——. 2002. A modest defence of school choice. *Journal of Philosophy of Education* 36:643–59.

——. 2005. *On Education*. London: Routledge.

Brock, D. W. 2006. Is a consensus possible on stem cell research? Moral and political obstacles. *Journal of Medical Ethics* 32:36–42.

Buchanan, A., D. W. Brock, N. Daniels, and D. Wikler. 2000. *From Chance to Choice*. Cambridge University Press.

Burley, J., and J. Harris. 1999. Human cloning and child welfare. *Journal of Medical Ethics* 25:108–13.

Caplan, A. (ed.). 1992a. *When Medicine Went Mad*. Totowa: Humana Press.

——. 1992b. Twenty years after: the legacy of the Tuskegee syphilis study. *Hastings Centre Report* 22:29–32.

Carroll, L. 1940. *The Complete Works of Lewis Carroll, Volume II, Through the Looking Glass*. London: The Nonesuch Press.

Chen, Y., Z. X. He, A. Liu, K. Wang, W. W. Mao, J. X. Chu, Y. Lu, Z. F. Fang, Y. T. Shi, Q. Z. Yang, D. Y. Chen, M. K. Wang, J. S. Li, S. L. Huang, X. Y. Kong, Y. Z. Shi, Z. Q. Wang, J. H. Xia, Z. G. Long, Z. G. Xue, W. X. Ding, and H. Z. Sheng. 2003. Embryonic stem cells generated by nuclear transfer of somatic nuclei into rabbit oocytes. *Cell Research* 13:251–63.

Christianson, A., C. P. Howson, and B. Modell (eds). 2006. March of Dimes global report on birth defects. White Plains, NY: The March of Dimes Birth Defects Foundation.

Council for International Organizations of Medical Sciences. 2002. International ethical guidelines for biomedical research involving human subjects. CIOMS report, Geneva.

Clarke, A., and F. Flinter. 1996. The genetic testing of children: a clinical perspective. In *The Troubled Helix* (ed. T. Marteau and M. Richards). Cambridge University Press.

Cohen, G. A. 1986. Self-ownership, world-ownership, and equality. In *Justice and Equality Here and Now* (ed. F. Lucash). Ithaca, NY: Cornell University Press.

——. 1989. On the currency of egalitarian justice. *Ethics* 99:906–44.

——. 1995. *Self-Ownership, Freedom, and Equality*. Cambridge University Press.

——. 2000. If you're an egalitarian, how come you're so rich? *The Journal of Ethics* 4:1–26.

Concetti, G. 1997. *L'embrione uno di noi*. Rome: Vivere In.

Crichton, M. 1999. *Timeline*. London: Random House.

Cornford, F. M. 1908. *Microcosmographia Academica*. London: Bowes and Bowes. (Reprinted 1966.)

Daniels, N. 1985. *Just Health Care*. Cambridge University Press.

——. 1996. *Justice and Justification*. Cambridge University Press.

——. 2000. Normal functioning and the treatment/enhancement distinction. *Cambridge Quarterly of Healthcare Ethics* 9:309–22.

——. 2007. Can anyone really be talking about ethically modifying human nature? In *Human Enhancement* (ed. J. Savulescu and N. Bostrom). Oxford University Press.

Davis, A. 1998. The status of anencephalic babies: should their bodies be used as donor banks? *Journal of Medical Ethics* 14:150–53.

——. 1994. All babies should be kept alive as far as possible. In *Principles of Health Care Ethics* (ed. R. Gillon), pp. 629–43. Wiley.

Dawkins, R. 2004. *The Devil's Chaplain*. London: Phoenix.

de Grey, A. D. N. J. 2004a. Escape velocity: why the prospect of extreme human life extension matters now. *PLoS Biology* 2:723–26.

—— (ed.). 2004b. Strategies for engineered negligible senescence: why genuine control of aging may be foreseeable. *Annals of the New York Academy of Sciences* 1019:1–597.

Devolder, K. 2006. What's in a name? Embryos, entities, and ANTities in the stem cell debate. *Journal of Medical Ethics* 32:43–48.

Devolder, K., and J. Harris. 2007. The ambiguity of the embryo: ethical inconsistency in the human embryonic stem cell debate. *Metaphilosophy* 38:153–69.

Devolder, K., and C. Ward. 2007. Rescuing human embryonic stem cell research: the possibility of embryo reconstitution after stem cell derivation. *Metaphilosophy* 38:245–63.

Dworkin, R. 1977. *Taking Rights Seriously*. London: Duckworth.

——. 1981a. What is equality? I. Equality of welfare. *Philosophy and Public Affairs* 10:185–246.

——. 1981b. What is equality? II. Equality of resources. *Philosophy and Public Affairs* 10:283–345.

——. 1993. *Life's Dominion*. London: HarperCollins.

——. 1996. *Freedom's Law*. Oxford University Press.

Edmonds, D. K., K. S. Lindsay, J. F. Miller, E. Williamson, and P. J. Wood. 1982. Early embryonic mortality in women. *Fertility and Sterility* 38:447–53.
Edwards, S. D. 2001. Prevention of disability on grounds of suffering. *Journal of Medical Ethics* 27:370–88.
Eliot, C. W. (ed.). 1909–14. *The Oath and Law of Hippocrates*.The Harvard Classics, volume XXXVIII, part 1. New York: Collier. (Available at www.bartleby.com/38/1/.)
Erin, C. A., and J. Harris. 1994. A monopsonistic market. In *The Social Consequences of Life & Death Under High Technology Medicine* (ed. I. Robinson), pp. 134–57. Manchester University Press.
——. 2003. An ethical market in human organs. *Journal of Medical Ethics* 29:137–38.
Feinberg, J. 1984. *The Moral Limits of the Criminal Law*. Oxford University Press.
Finnis, J. 1995a. A philosophical case against euthanasia. In *Euthanasia Examined: Ethical Clinical and Legal Perspectives* (ed. J. Keown). Cambridge University Press.
——. 1995b. The fragile case for euthanasia: a reply to John Harris. In *Euthanasia Examined: Ethical Clinical and Legal Perspectives* (ed. J. Keown). Cambridge University Press.
Foster, S. 2002. School choice and social justice: a response to Harry Brighouse. *Journal of Philosophy and Education* 36:291–308.
Fukuyama, F. 2002. *Our Posthuman Future*. London: Profile.
Gerami-Naini, B., O. V. Dovzhenko, M. Durning, F. H. Wegner, J. A. Thomson, and T. G. Golos. 2004. Trophoblast differentiation in embryoid bodies derived from human embryonic stem cells. *Endocrinology* 145:1,517–24.
Gilbert, W. S. 1956. *The Savoy Operas*. London: Macmillan.
Gillon, R. (ed.). 1994. *Principles of Health Care Ethics*. Wiley.
Giordano, S. 2007. Gender atypical organisation in children and adolescents: ethico-legal issues and a proposal for new guidelines. *International Journal for Children's Rights*, in press.
Glannon, W. 2002. Identity, prudential concern and extended lives. *Bioethics* 16:266–83.
Glover, J. 1977. *Causing Death and Saving Lives*. Harmondsworth: Penguin.
——. 1984. *What Sort of People Should There Be?* London: Pelican, Penguin Books.
——. 1999. *Humanity: A Moral History of The Twentieth Century*. London: Jonathan Cape.
——. 2006. *Choosing Children*. Oxford: Clarendon Press.
Green, R. M. 2001. *The Human Embryo Research Debates*. Oxford University Press.
Habermas, J. 2003. *The Future of Human Nature*. Cambridge: Polity Press.
Hari, J. 2005. History will judge the Pope far more harshly than the adoring crowds in Rome. *The Independent* (Comment), April 8, 2005.
Harris, J. 1980. *Violence and Responsibility*. London: Routledge and Kegan Paul.

Harris, J. 1983. *In vitro* fertilization: the ethical issues. *The Philosophical Quarterly* 33:217–37.
——. 1985. *The Value of Life*. London: Routledge.
——. 1987. QALYfying the value of life. *Journal of Medical Ethics* 13:117–23.
——. 1992. *Wonderwoman and Superman: The Ethics of Human Biotechnology*. Oxford University Press.
——. 1993. Is gene therapy a form of eugenics? *Bioethics* 7:179–89.
——. 1994. Not all babies should be kept alive as long as possible. In *Principles of Health Care Ethics* (ed. R. Gillon), pp. 643–57. Wiley.
——. 1996. What is the good of health care? *Bioethics* 10:269–92.
——. 1997a. Goodbye Dolly: the ethics of human cloning. *Journal of Medical Ethics* 23:353–60.
——. 1997b. What the principal objective of the NHS should *really* be. *The British Medical Journal* 314:669–72.
——. 1997c. The ethics of clinical research with cognitively impaired subjects. *The Italian Journal of Neurological Sciences* 18:9–15.
——. 1998. Rights and reproductive choice. In *The Future of Human Reproduction: Choice and Regulation* (ed. J. Harris and S. Holm). Oxford University Press.
——. 1999a. Ethical genetic research. *Jurimetrics: The Journal of Law, Science, and Policy* 401:77–92.
——. 1999b. The concept of the person and the value of life. *Kennedy Institute of Ethics Journal* 9:293–308.
——. 1999c. Justice and equal opportunities in health care. *Bioethics* 13:392–405.
——. 1999d. Clones, genes and human rights. In *The Genetic Revolution and Human Rights: The Amnesty Lectures* 1998 (ed. J. Burley). Oxford University Press.
——. 2000a. Is there a coherent social conception of disability? *Journal of Medical Ethics* 26:95–101.
——. 2000b. The moral difference between throwing a person at a trolley and throwing a trolley at a person: a reply to Frances Kamm. *Proceedings of the Aristotelian Society* (supplementary volume), 41–58.
——. 2000c. The welfare of the child. *Health Care Analysis* 8:27–34.
——. 2000d. Intimations of immortality. *Science* 288:59.
—— (ed.). 2001a. *Bioethics*. Oxford University Press.
——. 2001b. One principle and three fallacies of disability studies. *Journal of Medical Ethics* 27:383–88.
——. 2002a. Intimations of immortality: the ethics and justice of life extending therapies. In *Current Legal Problems* (ed. M. Freeman), pp. 65–95. Oxford University Press.
——. 2002b. The use of human embryonic stem cells in research and therapy. In *A Companion to Genetics: Philosophy and the Genetic Revolution* (ed. J. C. Burley and J. Harris). Oxford: Blackwell.

Harris, J. 2002c. Ethical issues in geriatric medicine. In *Textbook of Geriatric Medicine and Gerontology* (ed. R. Tallis and H. Fillett), 6th edn. London: Churchill Livingstone.
——. 2002d. Human beings, persons and conjoined twins: an ethical analysis of the judgement in *Re A*. *Medical Law Review* 9:221–36.
——. 2002e. Law and regulation of retained organs: the ethical issues. *Legal Studies* 22:527–49.
——. 2003a. Reproductive choice. In *Encyclopaedia of the Human Genome*. London: Nature Publishing Group Reference.
——. 2003b. Pro-life is anti-life: the problematic claims of pro-life positions in ethics. In *Scratching The Surface of Bioethics* (ed. M. Hayry and T. Takala). Amsterdam: Rodopi.
——. 2003c. Organ procurement: dead interests, living needs. *Journal of Medical Ethics* 29:130–35.
——. 2003d. Stem cells, sex and procreation. *Cambridge Quarterly of Healthcare Ethics* 12:353–72.
——. 2004a. *On Cloning*. London: Routledge.
——. 2004b. Immortal ethics. *Annals of the New York Academy of Sciences* 1019:527–34.
——. 2004c. Sexual reproduction is a survival lottery. *Cambridge Quarterly of Healthcare Ethics* 13:75–90.
——. 2005a. Reproductive liberty, disease and disability. *Reproductive Medicine Online* 10:13–16.
——. 2005b. No sex selection please—we're British! *Journal of Medical Ethics* 31:286–88.
——. 2005c. Scientific research is a moral duty. *Journal of Medical Ethics* 31:242–48.
——. 2005d. Sex selection and regulated hatred. *Journal of Medical Ethics* 31:291–94.
——. 2006. Mark Anthony or Macbeth: some problems concerning the dead and incompetent when it comes to consent. In *First Do No Harm* (ed. S. McLean). Aldershot: Ashgate Publishing.
——. Forthcoming. *The Safety of the People*. Oxford University Press.
Harris, J., and S. Holm. 1997. Why should doctors take risks? Professional responsibility and the assumption of risk. *The Journal of the Royal Society of Medicine* 90:625–29.
——. 2002. Extended lifespan and the paradox of precaution. *The Journal of Medicine and Philosophy* 27:355–69.
——. 2003a. Should we presume moral turpitude in our children? Small children and consent to medical research. *Theoretical Medicine* 24:121–29.
——. 2003b. Abortion. In *The Oxford Handbook of Practical Ethics* (ed. H. Lafollette). Oxford University Press.

Harris, J., and K. Keywood. 2001. Ignorance, information and autonomy. *Theoretical Medicine and Bioethics* 22:415–36.

Harris, J., and J. Sulston. 2004. Genetic equity. *Nature Reviews Genetics* 5:796–800.

Hart, H. L. A. 1955. Are there any natural rights? *The Philosophical Review* 64:175–91.

Hertig, A. T., and J. Rock. 1973. Searching for early fertilised human ova. *Gynecologic Investigation* 4:121–39.

Hobbes, T. 1960. *Leviathan* (ed. M. Oakeshot), part II, chapter 30, p. 219. Oxford: Basil Blackwell.

Holm, S. 1998. A life in the shadow: one reason why we should not clone humans. *Cambridge Quarterly of Healthcare Ethics* 7:160–62.

Human Fertilisation and Embryology Authority. 2003. Sex selection: options for regulation. A report on the HFEA's 2002–03 review of sex selection including a discussion of legislative and regulatory options. London: HFEA.

Hurlbut, W. B. 2005. Altered nuclear transfer as a morally acceptable means for the procurement of human embryonic stem cells. *The National Catholic Bioethics Quarterly* 5:145–51.

Jonas, H. 1972. Philosophical reflections on experimenting with human subjects. In *Experimentation with Human Subjects* (ed. P. A. Freund). London: Allen and Unwin.

——. 1992. The burden and blessing of mortality. *The Hastings Center Report* 22:34–40.

Jones, J. 2005. *Balliol College: A History*, 2nd edn. Oxford University Press (revised 2005).

Jones, R., K. Morris, and D. Nutt. 2005. Cognition enhancers. Report to Office of Science and Technology. Foresight Brain Science, Addiction and Drugs Project. (Available at www.foresight.gov.uk / Previous_Projects / Brain_Science_Addiction_and_Drugs / Reports_and_Publications / ScienceReviews / Cognition%20Enhancers.pdf; accessed March 1, 2006.)

Kamm, F. 1983. Killing and letting die: methodological and substantive issues. *Pacific Philosophical Quarterly* 64:297–312.

——. 1992. Non-consequentialism, the person as an end-in-itself, and the significance of status. *Philosophy and Public Affairs* 21:354–89.

Kass, L. R. 1997. The wisdom of repugnance. *The New Republic* 216:17–26.

——. 2001. L'Chaim and its limits: why not immortality? *First Things* 113:17–24.

——. 2003. Ageless bodies, happy souls: biotechnology and the pursuit of perfection. *The New Atlantis* 1:9–28. (Available at www.thenewatlantis.com / archive / 1 / TNA01-Kass.pdf.)

Kitcher, P. 1996. *The Lives to Come: The Genetic Revolution and Human Possibilities*. London: Simon & Schuster.

Koch, T. 2001. Disability and difference: balancing social and physical constructions. *Journal of Medical Ethics* 27:370–76.

Korn, D. 1998. Contribution of the human tissue archive to the advancement of medical knowledge and public health: a report to the National Bioethics Advisory Commission. In *Research Involving Human Biological Materials: Ethical Issues and Policy Guidance, Volume II, Commissioned Papers*. Rockville, MD: National Bioethics Advisory Commission.

Kuflik, A. 2007. The "future like ours" argument and human embryonic stem cell research. *Journal of Medical Ethics*, in press.

Landry, D. W., and H. A. Zucker. 2004. Embryonic death and the creation of human embryonic stem cells. *The Journal of Clinical Investigation* 114:1,184–86.

Lanza, R. P., J. B. Cibelli, and M. D. West. 1999a. Prospects for the use of nuclear transfer in human transplantation. *Nature Biotechnology* 17:1,171–74.

——. 1999b. Human therapeutic cloning. *Nature Medicine* 5:975–77.

Laurie, G. 1999. In defence of ignorance: genetic information and the right not to know. *European Journal of Health Law* 6:119–32.

Lee, K. S. 1994. Some Confucianist reflections on the concept of autonomous individual. *Journal of Chinese Philosophy* 21:49–59.

Le Grand, J. 2006. *Motivation, Agency, and Public Policy: Of Knights and Knaves, Pawns and Queens*. Oxford University Press.

Leridon, H. 1977. *Human Fertility: The Basic Components*. University of Chicago Press.

Li, X., W. Wei, J. Yong, Q. Jia, Y. Yu, and K. Di. 2005a. The genetic heterozygosity and fitness of tetraploid embryos and embryonic stem cells are crucial parameters influencing survival of mice derived from embryonic stem cells by tetraploid embryo aggregation. *Reproduction* 130:53–59.

Li, X., Y. Yu, W. Wei, J. Yong, J. Yang, J. You, X. Xiong, T. Qing, and H. Deng. 2005b. Simple and efficient production of mice derived from embryonic stem cells aggregated with tetraploid embryos. *Molecular Reproduction and Development* 71:154–58.

Liao, M. S. 2005. Rescuing human embryonic stem cell research: the blastocyst transfer method. *The American Journal of Bioethics* 5:8–16.

Locke, J. 1964. *An Essay Concerning Human Understanding*. Oxford University Press.

Lodge, D. (ed.). 1972. *Twentieth Century Literary Criticism: A Reader*. London: Longman.

Macklon, N. S., J. P. M. Geraedts, and B. C. J. M. Fauser. 2002. Conception to ongoing pregnancy: the "black box" of early pregnancy loss. *Human Reproduction Update* 8:333–43.

Maguire, E A., H. J. Spiers, C. D. Good, T. Hartley, R. S. J. Frackowiak, and N. Burgess. 2003. Navigation expertise and the human hippocampus: a structural brain imaging analysis. *Hippocampus* 13:208–17.

Marquis, D. 1989. Why abortion is immoral. *Journal of Philosophy* 86:183–202.

Marquis, D. 2005. Savulescu's objections to the future of value argument. *Journal of Medical Ethics* 31:119–22.

Marx, K. 1972. Theses on Feuerbach, no. XI. In *Marx and Engels* (ed. L. S. Fuer). London: Collins.

McBrearty, B. A., L. D. Clark, X. M. Zhang, E. P. Blankenhorn, and E. Heber-Katz. 1998. Genetic analysis of a mammalian wound-healing trait. *Proceedings of the National Academy of Sciences of the United States of America* 95:11,792–97.

McKie, J., J. Richardson, P. Singer, and H. Kuhse. 1998. *The Allocation of Health Care Resources: An Ethical Evaluation of the "QALY" Approach*. Aldershot: Ashgate Publishing.

McNeill, P. 1997. Paying people to participate in research: why not? *Bioethics* 11:390–96.

Mooney, D., and A. Mikos. 1999. Growing new organs. *Scientific American*, April 17, 1999.

Mill, J. S. 1962. On liberty. In *Utilitarianism* (ed. M. Warnock), chapter 1, p. 129. London: Collins Fontana.

Miller, P. (ed.). 2006. *Better Humans*. London: DEMOS.

Nagy, A., J. Rossant, R. Nagy, W. Abramow-Newerly, and J. C. Roder. 1993. Derivation of completely cell culture-derived mice from early-passage embryonic stem cells. *Proceedings of the National Academy of Sciences of the United States of America* 90:8,424–48.

Nayernia K., J. Nolte, H. W. Michelmann, J. H. Lee, K. Rathsack, N. Drusenheimer, A. Dev, G. Wulf, I. E. Ehrmann, D. J. Elliott, V. Okpanyi, U. Zechner, T. Haaf, A. Meinhardt, and W. Engel. 2006. In vitro-differentiated embryonic stem cells give rise to male gametes that can generate offspring mice. *Development Cell* 11:125–32.

Niwa, H., Y. Toyooka, D. Shimosato, D. Strumpf, K. Takahashi, R. Yagi, and J. Rossant. 2005. Interaction between Oct3/4 and Cdx2 determines trophectoderm differentiation. *Cell* 123:917–29.

Nozick, R. 1974. *Anarchy, State and Utopia.* Oxford: Basil Blackwell.

O'Neill, B. 2005. Did the Pope spread AIDS in Africa? *Spiked*, April 8, 2005. (Available at www.spiked-online.com / Articles / 0000000CA993.htm; accessed July 30, 2006.)

Orwell, G. 1970. *The Collected Essays, Journalism and Letters*, volume 1, p. 583. London: Penguin.

Overall, C. 2005. *Aging, Death, and Human Longevity: A Philosophical Inquiry*. University Presses of California.

Parfit, D. 1976. Rights, interests and possible people. In *Moral Problems in Medicine* (ed. S. Gorovitz). Englewood Cliffs, NJ: Prentice Hall.

——. 1984. *Reasons and Persons*. Oxford: Clarendon Press. (Reprinted 1987.)

Pedain, A. 2005. On cloning (book review). *Cambridge Law Journal* 64:507–9.

Pedersen, R. 1999. Embryonic stem cells for medicine. *Scientific American* 280:68–73.

Porter, R. 1999.*The Greatest Benefit to Mankind*, pp. 454–61, 652. London: Fontana.

President's Council on Bioethics. 2005. Alternative sources of human pluripotent stem cells. White Paper. Washington, DC: Government Printing Office.

Rachels, J. 1975. Active and passive euthanasia. *New England Journal of Medicine* 292:78–80.

——. 1979. Killing and letting people die of starvation. *Philosophy* 54:159–71.

Rawls, J. 1972. *A Theory of Justice*. Harvard University Press.

Raz, J. 1986. *The Morality of Freedom*. Oxford: Clarendon Press.

Redfern, M. (Chair). 2001. The Royal Liverpool Children's Inquiry Report. London: The Stationery Office.

Roberts, C. J., and C. R. Lowe. 1975. Where have all the conceptions gone? *Lancet* 1:498–99.

Robertson, J. A. 1994. *Children of Choice*. Princeton University Press.

Rose, S. 2006. Brain gain? In *Better Humans? The Politics of Human Enhancement and Life Extension* (ed. P. Miller and J. Wilsdon). London: DEMOS.

Russell, B. 1961. *Has Man a Future?* Harmondsworth: Penguin.

——. 2005. *Sceptical Essays*. London: Routledge Classics.

Sandel, M. J. 2002. What's wrong with enhancement. Presentation to The President's Council on Bioethics. (Available at www.bioethics.gov/background/sandelpaper.html.)

——. 2004. The case against perfection: what's wrong with designer children, bionic athletes, and genetic engineering. *Atlantic Monthly* 292(3):50–54, 56–60, 62.

Savulescu, J. 2002a. Deaf lesbians, "designer disability" and the future of medicine. *British Medical Journal* 325:771–73.

——. 2002b. Abortion, embryo destruction and the future of value argument. *Journal of Medical Ethics* 28:133–35.

——. 2004. Embryo research: are there any lessons from natural reproduction? *Cambridge Quarterly of Healthcare Ethics* 13:68–96.

Savulescu, J., and J. Harris. 2004a. The creation lottery: final lessons from natural reproduction: why those who accept natural reproduction should accept cloning and other Frankenstein reproductive technologies. *Cambridge Quarterly of Healthcare Ethics* 13:90–96.

——. 2004b. The great debates. *Cambridge Quarterly of Healthcare Ethics* 13:68–96.

Savulescu, J., B. Foddy, and M. Clayton. 2004. Why we should allow performance enhancing drugs in sport. *British Journal of Sports Medicine* 38:666–70.

Scanlon, T. 1989. A good start: reply to Roemer. *Boston Review* 20:819.

Scott, R. 2006. Choosing between possible lives: legal and ethical issues in preimplantation genetic diagnosis. *Oxford Journal of Legal Studies* 26:153–78.

Silver, L. M. 1998. *Remaking Eden*. Weidenfeld & Nicolson. (Paperback edn 1999. London: Phoenix Giant.)

——. 2000. Life and death in the 21st century. Interview in BBC *Horizon* TV program. (Broadcast January 4, 2000.)

Slevin, M. L., L. Stubbs, H. J. Plant, P. Wilson, W. M. Gregory, P. J. Armes, and S. M. Downer. 1990. Attitudes to chemotherapy: comparing views of cancer patients with those of doctors and the general public. *British Medical Journal* 300:1,458–60.

Spriggs, M. 2002. Lesbian couple create a child who is deaf like them. *Journal of Medical Ethics* 28:283.

Steiner, H. 1994. *An Essay on Rights*. Oxford: Blackwell.

Stevens, W. 1955. The man with the blue guitar. In *Collected Poems of Wallace Stevens*. London: Faber and Faber.

Stock, G. 2002. *Redesigning Humans*. Boston, MA: Houghton Mifflin. London: Profile Books.

Stoppard, T. 1972. *Jumpers*. London: Faber and Faber.

Sumner, L. W. 1987. *The Moral Foundation of Rights*. Oxford: Clarendon Press.

Swift, A. 2003. *How Not to Be a Hypocrite: School Choice for The Morally Perplexed Parent*. London: Routledge Falmer.

——. 2004a. The morality of school choice. *Theory and Research in Education* 2:7–21.

——. 2004b. The morality of school choice reconsidered: a response. *Theory and Research in Education* 2:323–42.

Takahashi, K., and S. Yamanaka. 2006. Induction of pluripotent stem cells from mouse embryonic and adult fibroblast cultures by defined factors. *Cell* 126:663–76.

Temkin, L. 1993. *Inequality*. Oxford University Press.

——. 2003. Egalitarianism defended. *Ethics* 113:764–82.

——. 2004. Thinking about the needy: a reprise. *The Journal of Ethics* 8:409–58.

Thomson, J. A., J. Itskovitz-Eldor, S. S. Shapiro, M. A. Waknitz, J. J. Swiergiel, V. S. Marshall, and J. M. Jones. 1998. Embryonic stem cell lines derived from human blastocysts. *Science* 202:1,145–47.

Thomson, J. J. 1971. A defense of abortion. *Philosophy & Public Affairs* 1:47–66.

Tolstoy, L. 1945. *Anna Karenin* (transl. R. Edmonds), part I. Harmondsworth: Penguin.

Tomasi de Lampedusa, G. 2005. *The Leopard*, p. 27. London: Vintage, Random House. (Transl. from *Il Gattopardo*, Universale Economica Feltrinelli. Ottantaseiesima edizione agosto 2005. Edizione conforme al manioscritto del 1957.)

Tooley, J. 2003. Why Harry Brighouse is nearly right about the privatisation of education. *Journal of Philosophy and Education* 37:427–47.

Toynbee, P. 2005. Not in my name. *The Guardian* (Comment), April 8, 2005.

Tsai, D. F. C. 2005. The bioethical principles and Confucius' moral philosophy. *Journal of Medical Ethics* 31:159–63.

UNESCO. 1997. Universal declaration on the human genome and human rights. December 3, 1997.

Waldron, J. 1988. *The Right to Private Property*. Oxford: Clarendon Press.

Weiss, K. M. 1985. The biology of ageing and the quality of later life. In *Ageing 2000: Our Health Care Destiny* (ed. C. Gaitz and T. Samorajski). Springer.

Weinrich, S. L., R. Pruzan, L. B. Ma, M. Ouellette, V. M. Tesmer, S. E. Holt, A. G. Bodnar, S. Lichtsteiner, N. W. Kim,, J. B. Trager, R. D. Taylor, R. Carlos, W. H. Andrews, W. E. Wright, J. W. Shay, C. B. Harley, and G. B. Morin. 1997. Reconstitution of human telomerase with the template RNA component hTR and the catalytic protein subunit hTRT. *Nature Genetics* 17:498–502.

Wilkinson, M., and A. Moore. 1997. Inducement in research. *Bioethics* 11:373–89.

Wilmut, I., A. E. Schnieke, J. McWhir, A. J. Kind, and K. H. S. Campbell. 1997. Viable offspring derived from fetal and adult mammalian cells. *Nature* 385:810–13.

Wittgenstein, L. 1966. *Tractatus-Logico Philosophicus* (transl. D. F. Pears and B. F. McGuiness). London: Routledge and Kegan Paul.

World Medical Association. 1964. World Medical Association Declaration of Helsinki: Ethical Principles for Medical Research Involving Human Subjects. Amended by the 52nd General Assembly, Edinburgh, Scotland, October 2000. (Available at www.wma.net/e/policy/pdf/17c.pdf.)

Wrong, M. 2005. Blood of innocents on his hands. *New Statesman*, April 11, 2005.

| Index